Management of Engineering Projects

Other Macmillan titles of interest to Mechanical Engineers

Mechanical Reliability, second edition
    A. D. S. Carter

Elementary Engineering Mechanics
    G. E. Drabble

Principles of Engineering Thermodynamics, second edition
    E. M. Goodger

Polymer Materials: An Introduction for Technologists and Scientists
    Christopher Hall

Introduction to Enginering Materials, second edition
    V. John

Strength of Materials, third edition
    G. H. Ryder

Mechanics of Machines
    G. H. Ryder and M. D. Bennett

Engineering Heat Transfer, second edition
    J. R. Simonson

Introduction to Internal Combustion Engines
    Richard Stone

Essential Solid Mechanics—Theory, worked examples and problems
    B. W. Young

Other Macmillan titles of interest to Civil Engineers

An Introduction to Engineering Fluid Mechanics
    J. A. Fox

Prestressed Concrete Design by Computer
    R. Hulse and W. H. Mosley

Reinforced Concrete Design by Computer
    R. Hulse and W. H. Mosley

Civil Engineering Materials, third edition
    edited by N. Jackson

Reinforced Concrete Design, third edition
    W. H. Mosley and J. H. Bungey

Microcomputer Applications in Structural Engineering
    W. H. Mosley and W. J. Spencer

Strength of Materials, third edition
    G. H. Ryder

Surveying for Engineers
    J. Uren and W. F. Price

Management of Engineering Projects

Edited by
Richard Stone

Brunel University,
Uxbridge, Middlesex

MACMILLAN
EDUCATION

First published 1988

Published by
MACMILLAN EDUCATION LTD
Houndmills, Basingstoke, Hampshire RG21 2XS
and London
Companies and representatives
throughout the world

Printed in China

British Library Cataloguing in Publication Data
Management of engineering projects.
    1. Engineering—Management
    I. Stone, Richard, *1955*
    620' .0068'4   TA190

    ISBN 0-333-40958-2
    ISBN 0-333-40959-0 Pbk

Contents

Preface

A project is by definition a transient phase that contrasts with the steady state operations more frequently found in industry.

For the successful completion of an engineering project, it is not sufficient to have a sound technical input. Business and project skills are also necessary, and these form the subject of this book. *Management of Engineering Projects* is divided into three parts:

Part I: Principles, Techniques and Management
Part II: Legal Aspects
Part III: Case Studies

The first three chapters constitute Part I, with Chapter 1 providing definitions of the phases in a project (Definition, Justification and Execution) and their associated financial aspects. Chapter 2 covers the project skills of estimation, planning, monitoring and control—an area where the integration of time and financial management is of the highest importance. Project management is discussed in Chapter 3, where the different types of project management structure are explained, along with the way that the structure can change during the project life cycle. The equally important subject of teamwork is also covered.

Part II of *Management of Engineering Projects* is devoted to the legal and related aspects of projects. The legal areas covered are contract law, and health and safety legislation. As greater use has been made of the project approach, so has greater use been made of contractors. For this reason Chapter 4 is devoted to contract types and contract law. Health and safety legislation now has a much greater significance, and despite its importance in all areas of employment, it is often omitted from degree and other courses. Chapter 5 also covers the related subject of engineering risk assessment and control.

The case studies in Part III are of two types: the first two are of a general nature, while the second two consider specific projects. Chapter 6 takes a general look at problems associated with establishing a new facility, and provides checklists to help ensure that a thorough approach is taken; Chapter 7 describes computer projects and some of their particular problems. The case studies in Chapters 8 and 9 illustrate the project techniques described earlier in the book, and they also provide insights into CAD/CAM and the subject of quality.

Discussion questions are placed at the end of each chapter. These can be used in seminars or by individual readers as a check on their understanding of what they have read. Many terms are defined in the text, and these are highlighted by an *italic* typeface. The terms that have been defined in the text are also indicated in the index by a **bold** typeface; this enables the index to be used as a glossary.

In the text no significance should be attached to the use of 'he' or 'his'; the alternatives of 'she' and 'hers' should be taken as implicit alternatives.

This book has been developed from material used in a final year course in the Special Engineering Programme at Brunel University. Some of the material has also been used for a short course on Project Management for engineers, consequently the book is aimed at engineering students and graduates. While the book can be used with students at any level, students who have received some industrial training and management teaching will probably find it easiest to appreciate the material. In order to keep the book to a readable and usable length, some material has only been presented in outline. However, sources of further information are listed in the Bibliography, as well as a list of references.

In the same way that a project is a multi-disciplinary activity, so has been the writing of this book. The Notes on Contributors indicate the backgrounds of the authors, and their contribution is obvious. Less obvious are the contributions from the numerous engineers and industrialists who have been willing to talk to me about projects. I would like to thank all the people who have contributed directly and indirectly to this book. Finally, I must thank Jenny Shaw, Peter Murby and John Winkler of Macmillan Education for their support in this project; special thanks are also due to Dennis Radford who read and commented usefully on the entire manuscript.

1988   Richard Stone

Notes on Contributors

David Birchall as Director of Graduate Studies at the Management College, Henley, is currently the Course Director of the Brunel University/Henley MSc in Project Management. He has had extensive experience of project management in construction, research and development, and in training/education. His publications include *Tomorrow's Office Today* published by Business Books, and many articles on work organisation, management development and project management.
(Author of Chapter 3)

Jonathan Hooker graduated from Loughborough University in Production Engineering and Management in 1979. He did two years of project engineering as described in the case study before becoming an assembly foreman, production control supervisor and finally shift supervisor. He joined Brunel University in 1984. He now works for PA Consulting Group.
(Author of Chapter 9)

Christopher Hudson is a Lecturer in Electronics and Computing in the Department of Engineering and Management Systems at Brunel University. Prior to this he was a Lecturer in the Computer Science Department at Queen Mary College (University of London), and before that he spent some 12 years in the electronics and computing industry. The industrial work was concerned with the design, implementation and management of computerised systems and projects for industry. He spent several years in the Computer Systems Group of Digital Equipment Company Ltd, where he project-managed many computer projects. His research interests are in the application of real-time systems, speech communications and aids for the disabled.
(Author of Chapter 7)

Martin Partington is a Professor of Law at Bristol University. He has researched widely into many areas of applied public law, including social welfare, social security and the regulation of the housing market. His contribution in this volume is an extension of that interest. For a number of years he has given lectures on law (including health and safety) to engineering students.
(Author of sections 5.2–5.7)

Mark Pearson graduated from the Special Engineering Programme at Brunel University in 1982. His first appointment was as a Project Engineer within the Manufacturing Engineering Department of a specialist engineering company. He became involved with numerous engineering projects, including shopfloor data collection and the capital justification for a machine tool and a computer aided design facility. He joined a leading CAD/CAM vendor in 1983 as an Applications Engineer, and more recently he moved into a sales role where he has responsibility for the Southern District operation.
(Author of Chapter 8)

Richard Stone is a Lecturer in the Department of Manufacturing and Engineering Systems at Brunel University. His interests and experience have covered a range of engineering projects; most recently this has taken the form of research projects. His involvement with Project Engineering takes the form of running a final year option for the Special Engineering Programme at Brunel and contributions to short courses, including some tailored to specific companies.

Geoffrey Woodroffe is a solicitor and a Senior Lecturer in Law and Director of the Centre for Consumer Law Research at Brunel University. He is a consultant for the National Consumer Council, the Consumers' Association and the EEC Commission. He has lectured, written and broadcast extensively on many aspects of contract law; he is the author of several books.
(Author of sections 4.3–4.8)

Notation

a_i	output from location i
AD	activity duration
ATE	automatic test equipment
b_j	requirement at location j
BOM	bill of materials
C_{ij}	transport cost from location i to j
CAD/CAM	computer aided design, computer aided manufacture
CNC	computer numerical control
CPA	critical path analysis
DCF	discounted cash flow
DOI	Department of Industry
ECGD	Export Credits Guarantee Department (part of the Department of Trade and Industry)
EEC	European Economic Community
EPROM	erasable programmable read only memory
ESD	earliest start date
FOC	Fire Officers Committee
FPA	Fire Prevention Association
HSC	Health and Safety Commission
HSE	Health and Safety Executive
HSWA	Health and Safety at Work Act
i	interest rate
IRR	internal rate of return
LFD	latest finish date

LPG	liquefied petroleum gas
m	learning constant
MAAPICS	MAnufacturing And Production Inventory Control System
n	number of sites, occurrences, years
NPV	net present value
PB	payback
PC	personal computer
PCB	printed circuit board
PERT	programme evaluation and review technique
PFBC	pressurized fluidized-bed combustor
PV(F)	present value (factor)
Q_i	quantity to be transported from existing site i
r	gearing ratio (equity finance/debt finance)
RAM	random access memory
t	time
T_i	transport cost per unit quantity
TC	total cost
TF	total float
TTC	tender to contract (scheme run by the ECGD)
UCTA	Unfair Contract Terms Act
USM	Unlisted Securities Market
VDU	visual display unit
X_{ij}	quantity to be transported from location i to j

Suffices

e	estimated
m	most likely
n	nth occurrence
o	optimistic
p	pessimistic

Part I

Principles, Techniques and Management

Chapter 1
Introduction

1.1 THE BACKGROUND TO PROJECT ENGINEERING

Engineering operations are becoming increasingly complex as a consequence of improving technology. This very often leads to a project approach, when a new product or process is being developed and introduced, in order to ensure that appropriate skills are available at all stages of the project. Project work requires not just a technical input, but also the correct attention to planning and teamwork.

Projects can take many forms, ranging from small-scale research and development activities through to major civil engineering or constructional projects, such as motorways and oil rigs. This range also encompasses projects in areas as diverse as chemical manufacture, computing (both hardware and software), and the manufacture of engineering and non-engineering goods for the industrial and consumer markets. Since project engineers can be involved at all stages of a project, it is not surprising that the number of project engineers is increasing, along with the importance of their role.

From the foregoing, a *project* can be defined here as a non-repetitive engineering activity of any size, from any aspect of industry; projects will frequently require a multi-disciplinary solution. The people involved with a project will have different roles, and this is reflected in their job title. In a small project, the following roles may all be taken by the Project Engineer:

Project Manager—the person who has overall responsibility for the success of a particular project, both from a technical and a business point of view, and is the customer's contact.
Project Leader—the person who has technical responsibility for a project and who directs the work of those working on the project. For small

3

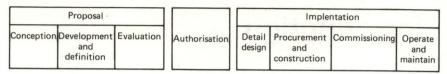

Proposal			Authorisation	Implentation			
Conception	Development and definition	Evaluation		Detail design	Procurement and construction	Commissioning	Operate and maintain

Figure 1.1   Main stages associated with a typical project

projects, the project manager and project leader may well be the same person but, for larger projects, the project manager will usually leave the day-to-day running of the project to the project leader, and just take on the business and customer side of the project.

Project Engineer—the person responsible for the design, development and construction of a particular aspect of a project. Any project may have several project engineers working on it, under the direction of the project leader.

Technician—the person responsible for the construction of parts of a project and possibly testing. The technical administration of drawings, part numbers, documentation etc. may also be under the jurisdiction of the technician.

The division of responsibilities between the roles, and indeed the titles of the roles, will vary between organisations and projects. For example, in a computer project the Programmer would be the counterpart of the Project Engineer.

Despite the widely different nature of many projects, the underlying principles of the project approach provide a unifying theme. The stages associated with a typical project are shown in figure 1.1. The authorisation or decision to proceed always defines the important difference between dealing with a proposal and the implementation of a project. The authorisation does not solely depend on technical merit, commercial merit can often be much more significant. Consequently engineers need an understanding of finance, marketing and the other commercial aspects, if they are to make a full contribution to their project. The best way of avoiding failed projects is not to have started them; thus rigorous project selection and evaluation procedures are essential. Without proper planning and control, the implementation stage in figure 1.1 can quickly degenerate, leading to under-achievement, over-spending, and even failure of the project (figure 1.2).

Initial enthusiasm	Disillusionment	Search for the guilty	Punishment of the innocent	Praise of the unworthy

Figure 1.2   Stages in a mismanaged project

1.2 PROJECT IDENTIFICATION AND DEVELOPMENT

As shown in figure 1.1, the first phase in any project is the *proposal* or *project identification*, where an outline of the project is formulated. The sources of a project can be as varied as the project itself; for example, *project conception*, the origination of the idea, can come from market surveys, new safety requirements, new techniques, new products etc. Of these areas, it is probably marketing that is least well understood by engineers.

The role of marketing in projects should not be underestimated; there is an important difference between Industrial Marketing and Consumer Marketing. Consumer marketing is often concerned with a product image, and persuading consumers that 'this powder washes whiter'. Perhaps surprisingly, the marketing costs can exceed the engineering costs in projects for manufacturing consumer goods. Marketing includes three quite distinct areas: strategic marketing, market analysis and marketing management. These are explained below:

> *strategic marketing*—identifying the company's strength and long-term aims

> *market analysis*—identifying product gaps, assessing the market reaction, analysing the competition

> *market management*—providing a back-up to sales through advertising etc.

As with any form of forecasting the predictions can only be estimates, but co-ordination with marketing in a project is essential to ensure that the forecast demand can be met. Any flexibility in the project is obviously an advantage.

The second stage in the proposal phase is the *development and definition of the project*. This is inevitably an iterative activity that involves both project evaluation and conception. As the idea of a project is developed, some preliminary estimates need to be made of the potential costs and benefits; this can then lead to the modification of the project concept.

As an example of a major project, consider the Pressurized Fluidized-Bed Combustor (PFBC) shown in figure 1.3, Pillai (1983). This complex plant offers a compact, efficient and flexible power generation system that should be capable of using low quality coal. Cost estimates for the various stages in the project are:

	£
feasibility study	10 000
detailed feasibility study	500 000
design costs	5–10 million
manufacture	300 million

While the cost of feasibility studies may appear high, they are still less than

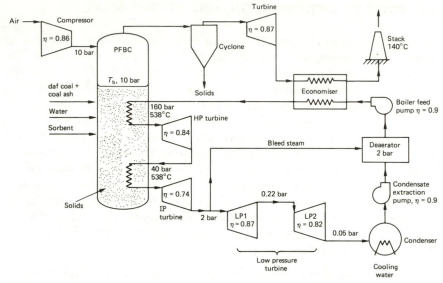

Figure 1.3   Example of a complex project—a Pressurized Fluidized-Bed Combustor. [Reprinted by permission of the Council of the Institution of Mechanical Engineers from K. K. Pillai (1983), 'Combined cycle power plant utilizing pressurized fluidized-bed combustors', Paper C72/83, Combustion in Engineering, Vol. II (I. Mech. E. Conference publications)]

1 per cent of the total cost. Also, by having two levels of feasibility study before any decision to proceed, the financial risk from starting an ill-conceived project is minimised. The way these figures have been presented also illustrates another important aspect of projects, namely estimating. The presence of only one significant figure implies an accuracy that may be no better than within an order of magnitude.

Despite the very high capital costs, larger (or more complex) operations should lead to economies of scale, and this can be illustrated by the *Boston experience curves*. The Boston Consulting Group has analysed a range of products and industries, and concluded that (all other things being equal) the real cost of a product falls by 20-25 per cent as the output doubles. Notable exceptions are caused by sudden cost increases for raw materials (such as oil in the 1970s), or conversely the introduction of new technology (such as integrated circuits replacing discrete components in electronic equipment).

Some Boston experience curves are shown in figure 1.4 for different goods, and the points of trade recession and recovery can also be identified. The selling price has to be made more competitive in times of recession to maintain the product turnover, while during recovery demand increases and the prices rise. The use of experience curves can be useful in deciding the size of a new production facility.

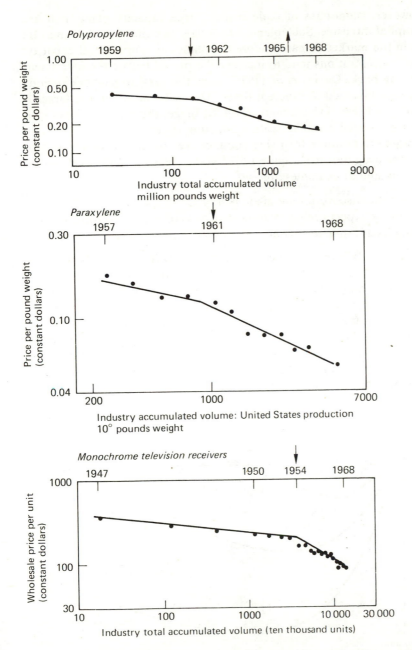

Figure 1.4 *Examples of Boston experience curves (points of trade recession (↓) or recovery (↑) [Reproduced, with permission, from Perspectives on Science, © 1970, The Boston Consulting Group, Inc.]*

However, economies of scale require large capacity plants that are highly capital intensive. Such plants take longer to build, and there may be a delay in the market demand before the plant can work at full capacity. All these considerations tend to increase the period before the investment starts to pay back. Davies *et al.* (1976) show the varying payback periods for polymers invented at different times. The later the plant, the larger it needs to be to break into the market, yet the bigger the risk associated with a process not tested on small-scale production (figure 1.5).

A dramatic example for rising entrance fees for new products is provided by the drug industry (figure 1.6). Two key reasons in the explanation for this and similar examples are:

(1) The simple discoveries are made first;
(2) consumers expect ever-increasing standards of product testing, and the testing itself tends to become more complex.

The risk of failure associated with a project will also be a function of the resources devoted to the project, and the degree of innovation (figure 1.7). Initially, additional resources (human and financial) will reduce the risk by ensuring a better design and planning, but as always the law of diminishing returns applies. Beyond a certain point the risk of failure increases; the solution can become over-engineered and ultimately too innovative. As is often the case, engineers should aim at 'fitness for purpose' rather than 'engineering excellence'. When the project proposal has finished being defined and developed, the project needs to be evaluated.

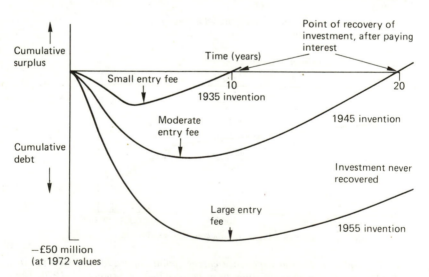

Figure 1.5 Cash flows of typical polymers invented in 1935, 1945, and 1955 [Reproduced, with permission, from D. Davies, T. Bamfield and R. Steahan (1976), The Humane Technologist, Oxford University Press, Oxford]

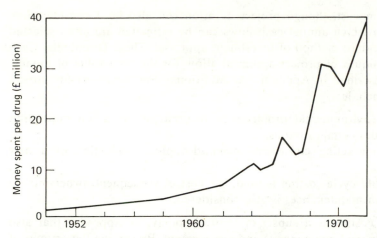

Figure 1.6   Escalating entrance fees for new drugs [Reproduced, with permission, from A. Spinks (1974), 'Escalating entrance fees for new drugs', ICI Magazine, Vol. 52, No. 411, © ICI Magazine]

1.3   PROJECT EVALUATION

By implication, the projects discussed so far are aimed at producing some financial return; obviously this is not always the case since *amenity projects*, such as new sewers or improved roads, a new canteen or washroom, cannot show a rate of return. Amenity projects are invariably chosen on the basis

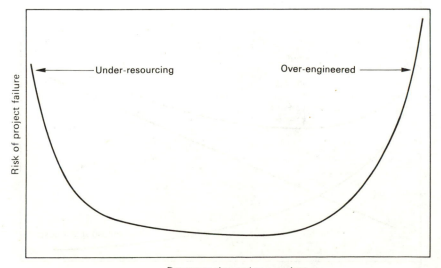

Figure 1.7   Risk of project failure through under-resourcing or over-engineering

of minimum cost, though project duration can also be a consideration. Projects on which annual cash flows can be estimated are often selected using a standard method of investment appraisal. These Discounted Cash Flow techniques of project appraisal allow for the time value of money, and are described in Appendix A. Additional considerations for project selection include:

(1) when environmental problems or insurmountable union opposition precludes certain options;
(2) when marketing constraints demand rapid introduction of a new product;
(3) when life cycle costing is used, the cost of subsequent ownership (in terms of maintenance) is also considered.

Life cycle costing is a subset of terotechnology, an approach that also includes planned maintenance and replacement. By varying the amount of maintenance the total cost can be minimised (figure 1.8).

Almost by definition projects need finance, and the source of the finance can vary enormously. For projects originated by Central Government the sources of finance include:

(1) Revenue from taxation;
(2) Money lent to savings schemes, such as National Savings, Premium Bonds etc.;
(3) Profits from state owned concerns;

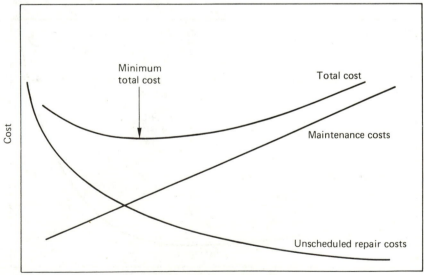

Figure 1.8 Maintenance optimisation

(4) Money borrowed from the international market;
(5) Money raised by selling National Assets.

Finance for Local Government projects can come from:

(1) Revenue from Rates;
(2) Money borrowed from investors (usually for a fixed period at a fixed interest rate);
(3) Money from Central Government.

The sources of finance for projects in private industry are perhaps the most diverse:

(1) Profits from previous years;
(2) Equity investment (often from private investors, institutions or other companies via the Stock Exchange);
(3) Debt finance (such as bank loans);
(4) Government subsidies and investment.

Profits from previous years can be a major source of finance for new projects, since it is not always possible to raise money on the Stock Exchange. Bank loans tend to be an expensive means of obtaining finance, and the Government can be fickle. However, grants from Central and Local Government for new facilities in particular areas, or for innovative techniques, are often significant, and these are discussed in Chapter 6.

Projects that will only recover their costs in the medium or long term (say longer than 2 years) require an appraisal technique that considers the time value of money. Two widely used techniques are evaluating the Internal Rate of Return and calculating the Net Present Value of a project. The *Internal Rate of Return* is calculated by treating the project as an investment, and deducing the interest rate generated by the project. The *Net Present Value* is the net profit or savings associated with a project (in current values) after a fixed period, which is the expected life of the facility produced by the project. Both these techniques use a Discounted Cash Flow technique, and they are both described in Appendix A. The inevitable problem with any such technique is the need for estimates of the project cost and duration, and the forecasting of the profits or savings associated with the new facility.

Unfortunately projects very often cost more than anticipated, and this necessitates reviewing the future of the project, and deciding whether or not to pull out. To continue, the potential profit by continuing must still be greater than the return from investing the outstanding project cost in either (1) another project within the same organisation or (2) the general financial market.

The compilation of annual accounts by organisations serves many useful purposes and is often a legal obligation. Nonetheless, it is most useful for steady state operations, and it can be a hindrance to engineering projects since it focuses attention on the short term. The revenue that a project might earn is inevitably an estimate based on sales or other forecasts. Also, there are very often intangible benefits associated with a particular project. New technology might enhance a company's image, new equipment or investment can improve an organisation's morale, or a competitor's investment may dictate a similar project.

Consequently, it can be misleading to compare projects, even on a basis of internal rate of return. Indeed, if the internal rates of return for projects were compared with guaranteed gains in the financial markets, it is likely that many projects would never be undertaken. However, if the proposers of a project believe sufficiently in their scheme, then they are likely to find the figures that provide the necessary justification.

After the project has been evaluated a decision has to be made whether to abandon the project, or to delay making a decision on the project, or to proceed with the project. If a project is authorised it is most likely to be for immediate action; this leads to the sequence of events described as project execution.

1.4 PROJECT EXECUTION

Project execution or implementation can include detail design, procurement, construction, commissioning, operation and maintenence; this is shown as the final sequence in figure 1.1. Planning monitoring and control are an implicit and essential aspect of this phase. This book is primarily concerned with the activities that occur during project execution, with most of the emphasis on the initial stages— that is, detail design, procurement and construction.

The activities in project execution do not have to be sequential, especially if there is a premium associated with rapid completion. For example, construction can start before the detail design work is completed, and commissioning can start before construction is complete. If this overlapping approach is used, it is almost inevitable that the project costs will be greater, since errors are more likely to occur and greater use is likely to be made of contractors. Almost inevitably an organisation has less control over contractors than its own work force. Furthermore, if the contracts are let for manufacture or construction before design is completed, then the contract is going to be let on less favourable terms. However, the savings in time can often be critical in a project's success.

Detail design is the 'fleshing out' of the ideas that have been produced by the development and definition stage of the project proposal. Unless the concepts are wrong, it is vital to accept the preliminary ideas and designs that have been produced in the preliminary phase. There is very rarely a single, optimum solution to an engineering problem, and change for the sake of change needs to be avoided. In other words, it is important to guard against the 'not invented here' syndrome, which is a prejudice for other peoples' ideas. Valuable time (and money) is often lost in a project by producing equally acceptable alternative designs. Some good design practices, such as the use of a top down approach, and brainstorming, are developed fully in Chapter 7, section 7.5.

The other problem that occurs in the detail design phase is when the project specifications are changed. The project specification needs to be frozen (once, and not several times), and this ought to occur prior to the authorisation or decision to proceed. The later into a project that a specification is changed, then the greater the costs attributable to the change. These costs are not restricted to increased time and expenditure, but can also include the lowering of morale and other intangible losses.

During procurement, installation and construction the rate of capital expenditure is very high, so that monitoring and control are essential. At this stage, any inadequacies in the evaluation or planning start to become apparent, and the success of the project can easily be jeopardised by over-spending and over-running. Perhaps most important of all is that items with a long lead time should be ordered as early as possible, to minimise delays occurring with procurement. It can also be worth over-ordering cheap items (such as fixings or electronic components), so as to ensure that these items are not the source of any delays.

Another critical stage is *commissioning and start-up*, when the first attempt is made to make the project hardware work. Wherever possible, sub-systems should be tested independently. Cost can also build up quickly through any delays in the start of useful production and the waste of raw materials. Some of the common problems associated with commissioning process plant are detailed by Kingsley *et al.* (1969); they include:

(1) errors in detail design (more likely than fundamental errors);
(2) automatic controls not working;
(3) organisational aspects.

The approach needed for testing and commissioning electronic and digital systems is described in Chapter 7, section 7.6.

Table 1.1 shows some of the sources of cost and time over-runs in each phase of project execution. It can be seen that the over-runs can originate at the project specification phase, or any subsequent phase. Furthermore, some actions or occurrences can have a cumulative effect. This means that the Definition, Design and Planning stages are often the most critical, despite their small contribution to the total project cost.

The profitability of any manufacturing operation is influenced very strongly by the reject or scrappage rate, or in the case of a process plant, the yield. This is especially the case in products with a low profit margin, for example consider the following product:

Table 1.1   Sources of Cost and Time Over-runs in Project Execution

---

1. Sources of over-runs occurring during the project specification
(a) Lack of previous experience
(b) Insufficient people with the necessary skills
(c) Inadequate time to prepare the specification
(d) Misunderstanding of the specification
(e) Agreement to terms and conditions that cannot be readily met
(f) Specification changes in the initial stages

2. Sources of over-runs occurring during design
(a) Incorrect estimating for time and/or costs
(b) Lack of resources, especially in the initial phases
(c) Inadequate control of contract labour
(d) Lack of project control
(e) Poor communications between the specification writer and the designers
(f) Specification changes during design
(g) Over-engineering
(h) Errors in design

3. Sources of over-runs occurring during procurement
(a) Placing orders late
(b) Failure to identify long lead items early enough
(c) Late delivery from suppliers
(d) Incorrect specification
(e) Incorrect supply
(f) Inability to supply equipment to original specification
(g) Goods lost, either in transit or at the factory

4. Sources of over-runs occurring during manufacture
(a) Shortage of suitable manpower
(b) Lack of documentation in both drawings and procedures
(c) Failure or non-conformance to specification of bought out parts
(d) Late delivery of components
(e) Correction of design faults
(f) Failure or shortages of manufacturing or test equipment
(g) Working space priority given elsewhere
(h) Strikes

% Yield	100	95	90
Sales income	100	95	90
Manufacturing costs	−60	−60	−60
Distribution costs	−30	−28.5	−27
Profit	10	6.5	3
% Profit on sales	10	6.8	3.3

If the product yield falls from 95 to 90 per cent, the fall in overall profit margin would reduce from 6.8 to 3.8 per cent, a fall that could have a disastrous effect on the pay back period of the project. Not surprisingly, quality and reliability are important considerations in any project. Indeed, many projects will have improvements in quality and reliability as their principal objective.

Organisational aspects can be critical, especially in the transition to start-up. Some of the questions that need to be resolved are:

(1) who is responsible for plant operation?
(2) at what stage is the plant handed over to the personnel responsible for production?
(3) who is responsible for the training of operators?

1.5 PROJECT AUDITING

A project does not end when construction or manufacture is completed. The project team should assist in the transition to use or operation, not least so that the members themselves can benefit from the experience of operation. This provides the most direct form of feedback, should there ever be a need for another similar project.

It is also worthwhile to have a formal review or audit of the project, and this should consider all aspects of the project—for example, the technical, managerial and commercial elements.

The criteria for assessing a project's success should include: was the project completed on time, within budget and to the technical specification?; is the finished project showing the required financial return?; and did the contractors benefit from the project?

Useful lessons can also be learned by examining other projects, and Morris (1987) discusses the following well-known projects:

Concorde This project was very successful technically, but there was very poor programme and financial control. Added to this, the problems of the high noise level and the sonic boom were not considered in advance, and this led to difficulties in obtaining take-off and landing rights.
Thames Barrier This is another technical success, although unforeseen constructional difficulties and inflation resulted in an over-run in cost and time. The scale of the project also led to managerial and labour relation problems.
Advanced Gas Cooled Nuclear Reactors (AGR) The second generation of AGRs are being completed essentially on time and within budget. This is a result of close contractor involvement and management (especially on site), and the technical lessons learned from the first generation of AGRs that suffered severe budget and schedule over-runs.
Giotto The European Space Agency launched Giotto in 1985, to intercept Halley's Comet 8 months later in March 1986. Like many of the European

Table 1.2   Some Prerequisites for Project Success

(a) The project must be clearly defined
(b) The amount of technical uncertainty and innovation should be minimised
(c) The project structure must be clearly defined, especially at the boundaries
(d) The project manager should not be encumbered with operational responsibilities unconnected with the project
(e) The planning and control systems should be as effective and as simple as possible
(f) Internal and external political support are needed
(g) The financial analysis should be thorough, and also consider the effects of delays and cost escalation from internal and external sources
(h) Good leadership, human relationships and labour relations are all needed
(i) There should be a minimum of changes to any specification
(j) It should be acknowledged that projects ultimately depend on people, none of whom is perfect.

Space Agency projects, this was a technical and managerial success. This is a result of strict project controls and leadership (which are especially important in the early stages) and through close attention to the high risk areas.

Advanced Passenger Train (APT)   By the time the technical difficulties for the APT had been resolved, the project had already attracted a lot of bad publicity. Potter (1987) explains that the management structure for the project also conflicted with existing British Rail functional departments, leading to much antagonism and little co-operation. Furthermore, during development it was difficult to accommodate the test schedules, and the maintenance depots were ill-equipped for servicing a high technology vehicle.

By reviewing these and similar projects, it is possible to identify the prerequisites for a successful project; some of the more important prerequisites are listed in table 1.2.

1.6   CONCLUSIONS

Already, it should have become clear that a successful project is not just reliant on a sound technical input, but also on business and project skills, this interdependence is shown in figure 1.9. The technical skills (whether engineering or scientific) are not discussed here, but the business and project skills that are needed in a project are the main subject of this book.

There is obviously an overlap between the business skills and the project skills, but they can be subdivided as follows:

business skills—management, finance and law

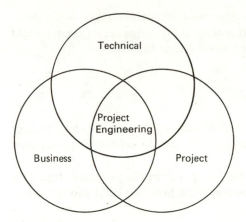

Figure 1.9   Attributes needed in Project Engineering

project skills—planning, estimating, monitoring and control

The main phases of a project, and the sequence of events in each phase, have already been shown in figure 1.1, and discussed in the subsequent sections.

The way to ensure a successful project is to allow sufficient iteration in the proposal phase, so as to produce a well-founded project proposal. This can then be authorised and subsequently implemented. During the implementation phase a minimum of deviation is desirable, and iteration should be avoided at all costs. The importance of the detail design stage is

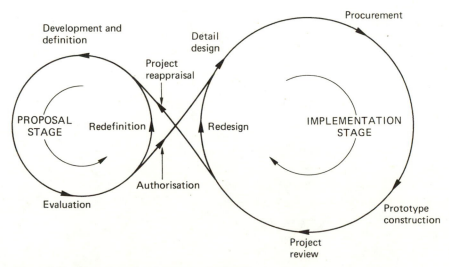

Figure 1.10   A project that is never completed

out of all proportion to its cost—the most important stage is the least expensive. If the project is not well managed the sequence of events might be as in figure 1.10, with the project never being finished.

1.7 DISCUSSION QUESTIONS

1. Define the roles of the Project Manager, Project Leader, Project Engineer and Technician in a project; discuss the relationships between the roles.
2. Detail the iterative processes that can occur in the proposal stage of a project, and the problems associated with both too little and too much iteration.
3. What are the reasons for the rising costs in projects for manufacturing products?
4. Describe the main stages in an engineering project, and explain why the project approach is widely used. In your opinion, what are the main causes of project failures?
5. Identify the sources of finance for engineering projects. Detail the methods for analysing a potential investment and the associated short-comings.
6. List the stages that occur after project authorisation. Identify the common sources of problems in projects, and elucidate the techniques to minimise the problems.
7. Why is the project approach being used in an ever-increasing amount? Comment on the advantages and disadvantages of large-scale operations.

Chapter 2
Project Timing and Financial Control

2.1 INTRODUCTION

For successful completion of any project, both the timing and costs have to be carefully planned, monitored and controlled. These two parameters are obviously interdependent, but none the less they are often treated separately. It is clearly fallacious to say that a project is on time simply because the expenditure matches the budget; this is most likely to be a warning of an over-spend and over-run. Equally, to monitor the progress solely by performance is also misleading since a project may be kept up to schedule by using overtime working and other costly resources.

The techniques presented in this chapter for planning project timing are in order of increasing complexity—always use the simplest technique that is consistent with the requirements. A network analysis might look more impressive than a bar chart, but it might end up confusing the originator as well as the project team. As with any system, it is normally easier to keep a project on course than to bring it back to schedule after a disruption. Thus good scheduling systems should have scope for comparing the actual achievement with the predictions.

Financial control should be an integral part of the project timing control. In order firstly to estimate the project timing and secondly to monitor the progress, the project will be broken down into self-contained units. Depending on the size and complexity of the project, the units might then be subdivided. The costs associated with the units and subdivisions can then be estimated, to provide a total cost and the spending rate as a function of time. The modifications that are needed when the project diverges from the schedule are discussed in section 2.3.3.

Computers have an obvious role in assisting the planning and monitoring of projects. The choice of software is going to be influenced by the computer that is available and the type of project that is being undertaken. A computer facilitates project monitoring, and if a project needs to be rescheduled then the additional work for the operator is minimal.

Computer aided design and manufacture (CAD/CAM) also has a significant contribution to make in projects. The design process can be speeded up, and certain checks are carried out automatically—for example, interference between moving parts, or between piping and a structure. Also parts lists can be generated automatically, and this could be coupled directly to a computer aided planning system. The more sophisticated analysis techniques that are feasible with computers reduce the need for prototypes, and this leads to direct savings in both cost and time. These and other advantages of CAD/CAM are discussed more fully in chapter 8 where the advantages of CAD/CAM are discussed as part of a particular CAD/CAM project.

Finally, good information cannot solve problems, but poor information can create them.

2.2   PROJECT TIMING

2.2.1   Introduction

The use of scheduling methods in engineering operations (whether some form of bar chart or network analysis) is covered by many books, for example, Wild (1984). However, with either batch or continuous production the scheduling requirements are somewhat simpler. By definition a project is not a recurrent activity, thus the time and cost associated with each activity is an estimate, and consequently there is an important need for information that monitors the progress of the project. Books on project management such as Lock (1977) and Harrison (1985) give comprehensive treatments of scheduling techniques, but with less emphasis on the project monitoring aspects, and the integration of financial control.

Before dealing with specific techniques it is worth remembering that there are two extreme types of project, though most fall somewhere between the two:

(1) known technology for a known environment;
(2) novel technology for a novel system.

An example of the first type of project is a bridge of conventional design that has to be built by a specific date. This is a project that uses known technology, so the bridge design and scheduling can be geared to the completion date by ensuring sufficient resources are available.

The second type of project is typical of many research activities, but even these can be constrained by using the best 'device' available by a specific date. However, great care is needed to avoid the effects of

Parkinson's 1st Law, namely 'Work expands so as to fill the time available for its completion' (Parkinson, 1981).

Before planning with units of weeks or months, remember that the available working man–days each week may vary owing to statutory and other holidays, sick leave, and through interference from external factors, such as the availability of the site, bad weather, or adjacent activities on existing production operations etc.

2.2.2 Bar Charts

Bar charts are long established and easy both to construct and to interpret, yet they are also very versatile. With a *bar chart*, each activity has a separate bar, the length of which corresponds to the activity duration. The horizontal position of the bar also identifies the start and finish times. Bar charts are sometimes referred to as *Gantt charts* after Henry Gantt who adapted them for monitoring progress, by using a separate bar to show achievement. A simple example is shown in figure 2.1, for the erection of a prefabricated garage.

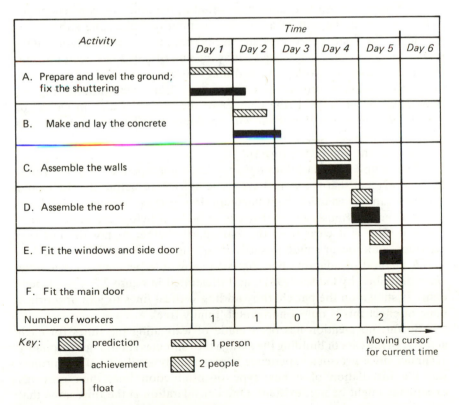

Figure 2.1 Bar chart for garage erection

It is self-evident that the foundations have to be prepared and the concrete plinth has to be laid and left to harden, before the structure can be built. This aspect can be undertaken by a single worker, while the prefabricated sections in the garage need two people because of the size and weight. In this particular bar chart the height of the bars reflects the number of people involved. Alternatively separate bars could be used for each worker, or type of worker if there needs to be a distinction between skilled, and semi-skilled workers.

In this example solid bars record the achievement, and this shows that at the end of day 5 the project should be complete, while in fact the main door still needs to be fitted. If completion by the end of day 5 was written into a contract with a penalty for late completion, then it would have been prudent to employ two people for the whole time so that the concrete base was finished by day 1. Ensuring even usage of equipment and personnel is an important aspect in project planning, and is referred to as *resource levelling*. Extra rows can be added to show the utilisation levels of personnel or key equipment. Alternatively the same information can be displayed as a histogram underneath the bar chart.

The bar chart can also show *float*—that is, the time that can accommodate over-run at the end of an activity without affecting the subsequent activities. The float is shown as an 'empty' bar; such a chart is sometimes referred to as a *chain chart*. The float at the end of day 2 has been utilised, possibly because the worker decided that two days would be taken for the job. Float obviously provides for unforeseen delays without affecting the subsequent programme. If the planner of the project is also in charge of the project execution, then float will no doubt also be incorporated into the key or critical activities (*critical activities* are those that dictate the timetable of the complete project).

This simple bar chart does not show the interdependence of activities. It is self-evident that the walls have to be assembled before the doors and windows can be installed, but if the main door runners are suspended from the roof then obviously the roof needs to be built before the main door is installed. Nor is it clear from this bar chart whether or not the last two activities could occur simultaneously if sufficient workers were available.

A method of adapting bar charts to show interpendence is the *stage chart* described by Oddey (1981), and illustrated in figure 2.2. Interdependence is shown on the bar chart by adding vertical lines to join apropriate bars; in effect this is now a horizontal 'family tree'.

So far, no mention has been made of estimating the times for each activity. In the case of building the garage the estimates should be reasonably accurate, since previous experience is presumably available. If the project was the installation of a new type of production line, then previous experience might be non-existent. One consideration is the time units that are used; units that are too small give an unwarranted impression of

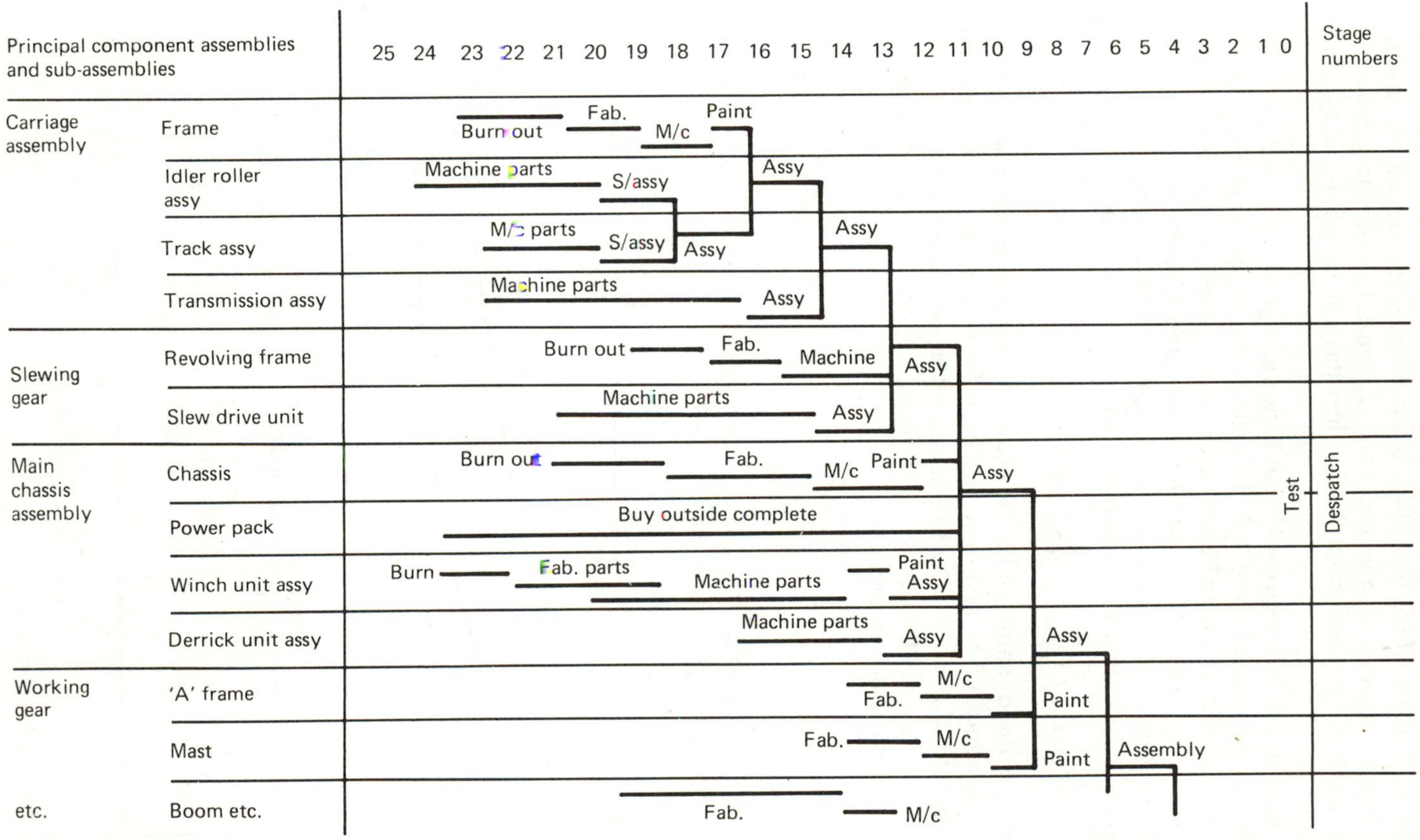

Stage numbers

25 24 23 22 21 20 19 18 17 16 15 14 13 12 11 10 9 8 7 6 5 4 3 2 1 0

Test   Despatch

Principal component assemblies and sub-assemblies

Carriage assembly
- Frame — Burn out, Fab., M/c, Paint, Assy
- Idler roller assy — Machine parts, S/assy, Assy
- Track assy — M/c parts, S/assy, Assy
- Transmission assy — Machine parts, Assy

Slewing gear
- Revolving frame — Burn out, Fab., Machine, Assy
- Slew drive unit — Machine parts, Assy

Main chassis assembly
- Chassis — Burn out, Fab., M/c, Paint, Assy
- Power pack — Buy outside complete
- Winch unit assy — Burn, Fab. parts, Machine parts, Paint, Assy
- Derrick unit assy — Machine parts, Assy

Working gear
- 'A' frame — Fab., M/c, Paint, Assembly
- Mast — Fab., M/c, Paint

etc.
- Boom etc. — Fab., M/c

Figure 2.2 Simple stage chart in excavator manufacture and assembly [Reprinted by permission of the Council of the Institution of Mechanical Engineers from J. Oddey (1981), 'Back to basics', CME, Vol. 28, No. 11]

accuracy. While it would be realistic to estimate cycle times on the production line in seconds or minutes, when it comes to estimating times for installation of the production facility then days or weeks would be more appropriate.

If an activity has been carried out a few times previously, then a *learning curve* may be useful (figure 2.3). This shows the exponential reduction in activity time that is said to be characteristic of some repetitive tasks. There is of course a time (t_r) below which the task cannot be completed, even on a repetitive basis. If t_n is the activity duration on the nth occurrence, then assuming the exponential redirection in activity duration:

$$t_n = t_r + (t_1 - t_r) \exp[-(n-1)/m]$$ (2.1)

where t_1 is the duration on the 1st occasion and m is a learning constant.

The values of m and t_r/t_1 can only be found by experience, and consequently the values of t_n will in general be only approximate.

In the case of PERT analysis (section 2.2.3) the estimated time (t_e) is based on a weighted mean:

$$t_e = \frac{t_0 + 4t_m + t_p}{6}$$ (2.2)

where t_0 is an optimistic time for an activity, t_m is the most likely time for an activity, t_p is a pessimistic time for an activity.

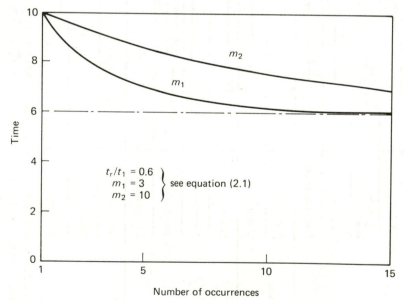

Figure 2.3 *Learning curves*

However, to allow for estimates frequently being optimistic, it may be better to use the following estimate:

$$t_e = \frac{t_0 + 3t_m + 2t_p}{6} \qquad (2.3)$$

Within reason, it is obviously better to obtain time estimates from people who will be responsible for an activity. Not only is the estimate likely to be most accurate from the people responsible for an activity, but they may also have some personal commitment subsequently to maintain the schedule. However, if a person's achievement is going to be compared with his prediction, that person is likely to give a conservative estimate of the duration. Before constructing a bar chart, the timescale has to be fixed—either days, weeks or months. Also, a choice needs to be made between 'project time' (day 1, 2, 3 etc.) or 'real time' (June 1, June 2 etc.). If real time is used with units of weeks, then they can be specified by the date at the beginning of each week or the week number, since it is now common for diaries to include the week number.

Aids to constructing bar charts include computer programs, wall charts and pro formas. Whether a wall chart uses a magnetic or peg system, it is best kept away from corridors and accidental knocks. While colours improve the impact of a wall chart, they make recording more difficult; furthermore some 8 per cent of the male population have congenitally defective colour vision. Again, pro formas are best restricted to black and while for ease of copying.

2.2.3 Network Analysis

Network analysis describes several planning techniques, of which *critical path analysis* (CPA) and *programme evaluation and review technique (PERT)* are the best known; both techniques were devised in the USA in the 1950s. The *arrow* or *network diagrams* used in CPA or PERT show the interdependence of activities, but unlike bar charts their length has no relation to the activity duration.

Networks are constructed by drawing arrows between numbered nodes in order to represent activities; conventionally time increases from left to right, and the node numbering is purely for ease of identification. The horizontal position can be made proportional to time, and horizontal bands can be used to segregate different types of activity (for example, Design, Procurement etc.).

Two aspects of vehicle servicing are shown in figure 2.4. The arrows show the sequence of events and the precedence, for example: before the engine oil can be replaced the new oil needs to be bought and the old oil has to be drained.

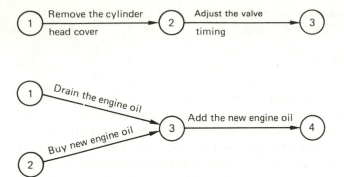

Figure 2.4   Simple networks for vehicle servicing

Some activities occur sequentially while others can occur in parallel; this is shown in figure 2.5 for a shelving project.

The activities shown by the solid arrows require a finite time; sometimes it is necessary to use a *dummy activity*—an activity of zero duration shown by a dashed line. For example:

(1) To avoid a parallel activity with the same consecutive beginning and end event. Figure 2.6a can be redrawn as figure 2.6b or 2.6c, though figure 2.6b would be preferable since activity G is less likely to be overlooked when shown this way.
(2) To show a one way dependence on activities. In figure 2.7a the final design of the power supply (H) depends on the design of the amplifier (D), while manufacture of the amplifier (G) does not depend on the preliminary design of the power supply (E). If the network had been drawn as in figure 2.7b, then the implication is that the amplifier manufacture (G) is dependent on the preliminary design of the power supply (E). The dummy activity removes this ambiguity.

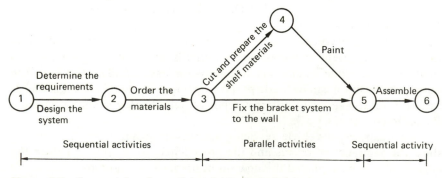

Figure 2.5   Sequential and parallel activities in a shelving project

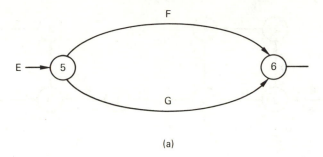

(a)

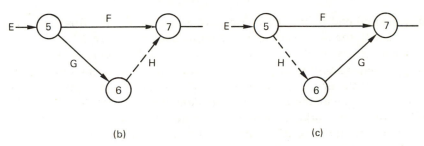

(b)   (c)

Figure 2.6   *Use of dummy activity to avoid ambiguity with parallel activities*

(3) To clarify a drawing. The networks shown in figure 2.8 are equivalent, but the dummy activity makes the representation easier; this is most likely to be necessary in complicated networks.

The purpose of constructing a network is to provide a framework for calculating and controlling the overall duration of a project. In order for this to be feasible, two elements have to be eliminated from any diagram: loops and 'dangles'. *Loops* imply the repetition of an activity sequence such as in figure 2.9a, and these need to be drawn in full (figure 2.9b). *Dangles*

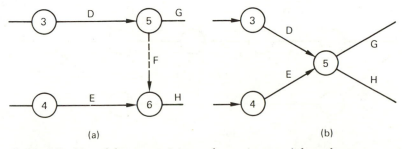

(a)   (b)

Figure 2.7   *Use of dummy activity to show a 'one-way' dependence*

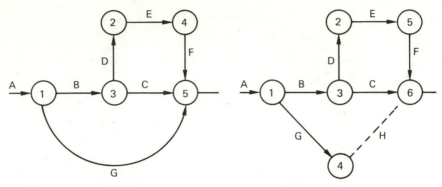

Figure 2.8   Use of dummy activity to simplify a diagram

are additional end points for an aspect of a project (figure 2.10a), and can be brought to the main ending by a dummy activity (figure 2.10b).

Each activity needs a time assigned to it, and the accuracy of the time estimates determines the utility of the network analysis. Using the network in figure 2.11, the first step is to label the activities, nodes, activity duration, and then to calculate the *earliest starting date (ESD)* for each activity. The ESD is found by working from zero time at the left of the diagram, and adding the activity times, the result being recorded adjacent to each node.

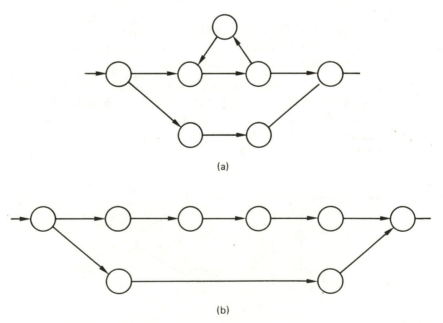

(a)

(b)

Figure 2.9   Elimination of loops in networks

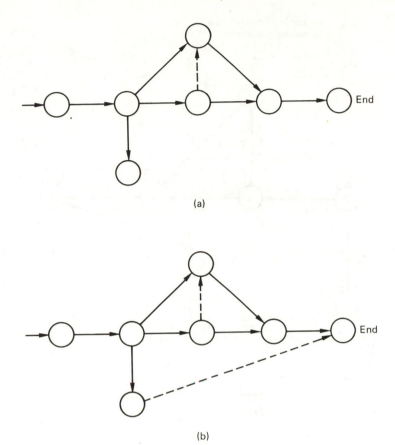

(a)

(b)

Figure 2.10   Elimination of dangles in networks

Conventions differ—the ESD may or may not be in a box, either above or below the node (figure 2.11). Where the activity following a node is dependent on completion of more than one activity, then the earliest starting date (ESD) is dictated by the slowest activity (highest ESD); this occurs at nodes 5 and 6 in figure 2.11.

The next step is to calculate the *latest finishing date (LFD)* for each activity if the project is to be completed within the time calculated, in this case 15 days. The LFD is calculated for each node by working from the end, and subtracting each activity duration. Again conventions differ, but in figure 2.12 each LFD is recorded beneath the relevant node. Where a node leads to more than one activity there will be more than one LFD, but obviously the lowest numerical value governs.

The difference between the earliest starting date (ESD) and the latest finishing date (LFD) for an activity gives the time available for the activity.

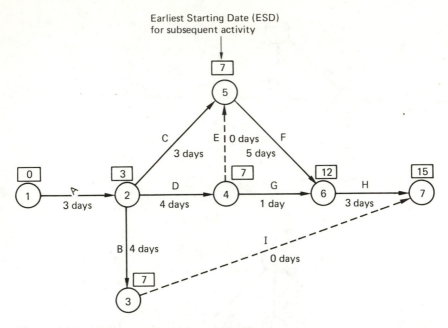

Figure 2.11 Earliest starting dates on a network diagram

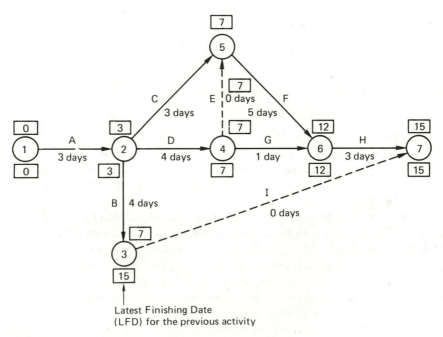

Figure 2.12 Latest finishing dates on a network diagram

The difference between the time available for an activity and the activity duration (AD) is the *total float* (*TF*), in other words:

$$TF = (LFD - ESD) - AD \qquad (2.4)$$

For example, the total float in activity G is 4 days. It is important to realise that if activity G was replaced by a sequence of activities then the total float on each activity in the sequence would also be four days, but (obviously) the total float is shared between these activities, and cannot be used more than once. This leads to definitions of three other types of float which are not necessarily positive:

(1) *Free float* (*early*) = Earliest starting date of succeeding event
  − Earliest finishing date for preceding event
  − Activity duration

(2) *Free float* (*late*) = Latest starting date of succeeding event
  − Latest finishing date for preceding event
  − Activity duration

(3) *Independent float* = Earliest starting date of succeeding event
  − Latest starting date for preceding event
  − Activity duration

Free float is not a shared float, in as much as the same float is not shown on different activities. Independent float is the free time available if the preceding and succeeding activities are as close together as possible; the use of independent float cannot affect other activities.

Where total float on an activity is zero, then the activity is critical, and the sequence of *critical activities* defines the *critical path*. In figure 2.13 the critical path is shown by bold lines. By definition, other activities have float associated with them, and if a critical activity over-runs then the floats on the other parallel activities will increase. If the total project duration needs to be reduced, then attention has first to be paid to the critical activities. However, reductions in the duration of critical activities may make other activities critical. For example, if the duration of activity D is reduced to 2 days then activity C becomes critical, and the network analysis will have to be repeated.

Network analysis also enables activities to be identified that are eligible for rescheduling in order to level resource utilisation, both within an individual project and in the context of parallel projects.

In this section the terms PERT and CPA have been avoided since they are two of the many different network methods. The terms are often used interchangeably; but CPA uses a simple time estimate while PERT uses the weighted mean estimate for activity duration given by equation (2.2).

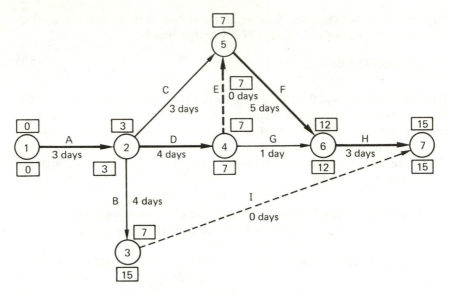

Figure 2.13   Critical path on a network diagram

2.2.4   Resource Levelling

If resources are to be used efficiently it is self-evident that their utilisation needs to be evened out or optimised. If the prefabricated garage discussed in section 2.2.2 was one of many to be built on a housing estate, then an efficient schedule can be obtained simply by inspection.

Referring back to figure 2.1, it can be seen that activities A and B require 1 person for 2 days, while activities C, D, E and F require 2 people for 2 days. By allowing the concrete plinth to cure for 2 days, a simple pattern emerges, in which a team of 3 workers would become fully occupied; this is shown in figure 2.14 for 7 garages.

The obvious drawback with this scheme is that the three workers are not fully utilised all the time; this of course would be less significant if 70 garages were being built rather than just 7. An alternative approach would be to build all the foundations and then erect all the garages. Assuming that the materials can be supplied at the appropriate rate, this may appear attractive. However, there would be less job variety, and the builder may be paid on the basis of garages completed, or the contract may specify certain staged completion dates.

A compromise that provides complete labour utilisation and a steady rate of completion is shown in figure 2.15 for the garages. This scheme also permits job variation within the group of 3 workers, except at the beginning and end. Also, since the garages are completed on a steady basis this

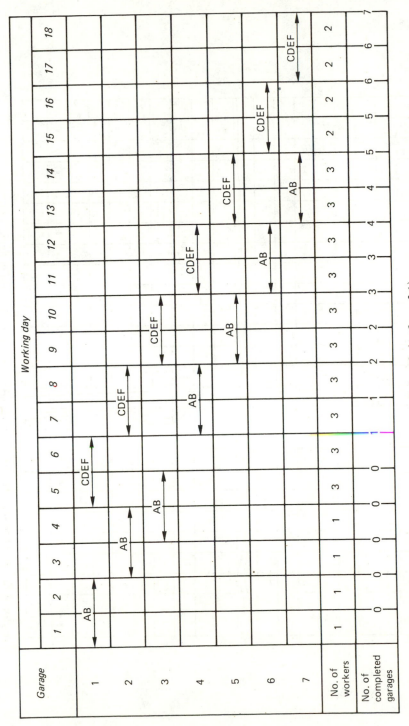

Figure 2.14 Building sequence for seven garages (Activities A–F identified in figure 2.1)

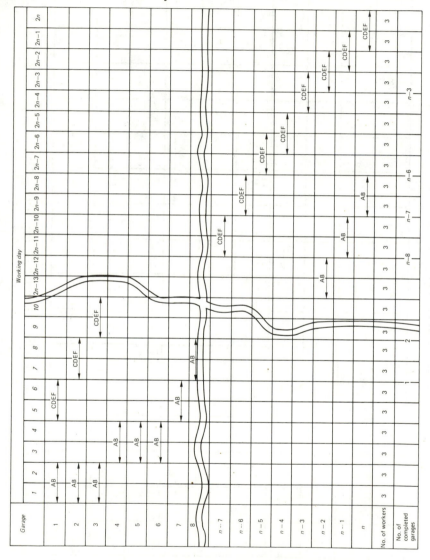

Figure 2.15   Building sequence for n garages (Activities A–F identified in figure 2.1)

provides better motivation for the workers, and a better cash flow for the business.

For more involved working patterns, a technique such as the *line of balance* is appropriate. In manufacture, the line of balance technique is used to ensure that the necessary sub-components for an assembly are manufactured in sufficient time to meet a prescribed production pattern; this application is discussed by Wild (1984). The line of balance can also be used to design for resource levelling and this is described in some detail for the construction industry by Oxley and Poskitt (1980). Since the line of balance is only appropriate to the production of repetitive units it will not be elaborated here.

2.2.5 Computer Based Systems and a Comparison between Bar Charts and Network Analysis

A comparison between a bar chart and a network diagram for the same project is shown in figure 2.16 (Lock, 1971). The advantages of a bar chart are that:

(1) it is readily intelligible;
(2) it provides a basis for an operational plan;
(3) it is easy to construct;
(4) it is capable of refinement—for example, as a stage chart

Unfortunately the bar chart is not sufficiently sophisticated for large projects, nor is it easy to update and modify when plans have to be changed.

Network analysis is most appropriate for large projects, but it then becomes time-consuming to calculate, and its complexity discourages updating, especially if manual methods are used. A widely held view is that if a project is large enough to need network analysis, then the network analysis will need a computer. Fortunately, a wide variety of software is available, and it is suitable for micros and PCs (personal computers) upwards. The computer program will also be able to check the logic in the network; it will be able to identify activities not connected to two events, more than one activity connected to the same events, loops and dangles. Errors that are plausible cannot of course be identified by a computer.

The programs should also be able to deal with multiple starts and finishes. Suitable means of displaying and outputting the information are of course essential. Bar charts can also be generated from computer programs, and there is no reason why a combination of both bar charts and network analysis cannot be used. A hierarchical approach will be necessary to provide sufficient detail with manageable sized charts. Some of the activities at each level are shown in table 2.1 for a project to produce a new car.

Job	Day number																								
	1	2	3	4	5	6	7	8	9	10	11	12	13	14	15	16	17	18	19	20	21	22	23	24	25
Design concept	▨																								
Design trolley	▨	▨	▨	▨																					
Buy trolley fittings						▨	▨	▨	▨	▨															
Design sub-frame					▨	▨																			
Design panels and door							▨	▨	▨	▨															
Buy all raw materials							▨	▨	▨	▨	▨	▨	▨	▨											
Buy door fittings and trims											▨	▨	▨	▨	▨	▨	▨	▨	▨	▨					
Make trolley parts															▨	▨	▨								
Make sub-frame parts															▨	▨	▨								
Weld and clean up trolley																		▨							
Weld and clean up sub-frame																		▨	▨						
Make door and panels															▨	▨	▨	▨	▨	▨					
Paint all parts																					▨	▨			
Drill and assemble																							▨	▨	

RESOURCE LOADING

	1	2	3	4	5	6	7	8	9	10	11	12	13	14	15	16	17	18	19	20	21	22	23	24	25
Design staff	1	1	1	1	1	1	1	1	1	1	0	0	0	0	0	0	0	0	0	0	0	0	0	0	0
Workshop staff	0	0	0	0	0	0	0	0	0	0	0	0	0	0	3	3	3	3	1	1	1	1	1	1	0

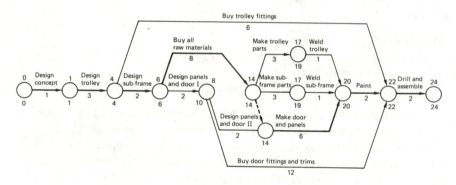

Figure 2.16   Comparison between bar chart and network diagram for the same project [Reprinted by permission of the Council of the Institution of Mechanical Engineers from D. Lock (1971), 'Scheduling industrial projects', CME, Vol. 18, No. 9]

The application of computers to project management is discussed at length by Harrison (1985). Computers help with the project planning and control once the project has commenced. At the planning stage computers can provide predictions of the equipment needs, the human resources and the financial needs. A computer can also enable the schedules to be optimised. Some of the constraints might be:

(1) to minimise the project duration;
(2) to ensure level usage of resources;
(3) to optimise the cash flow (by delaying major expenditure);
(4) to enable project execution with limited resources.

Table 2.1   A Hierarchial Approach to Project Control

Level	Activities (to include)
1. New car	Body styling Structural analysis Powertrain development Aerodynamic development Suspension and steering development Production planning etc.
2. Powertrain	Engine development Gearbox development Rear axle development Powertrain optimisation etc.
3. Diesel engine development	Basic engine design Ancillaries Fuel injection system Turbocharged derivative Cylinder head design etc.

etc.

During project execution the computer can be updated with the progress of activities and their costs. The computer would then be able to compare progress with the predictions, and re-optimise the schedule. Comparison between the cost predictions and the actual expenditure would also be available to provide warning of an over-spend. Some of the considerations needed in selecting a software package for project work are given in table 2.2.

Table 2.2   Checklist for selecting Network Analysis Software

1. Can it be used for monitoring and control, or just planning?
2. Is it menu driven?
3. What format does the output have (bar charts and/or networks), and how is the critical path identified?
4. What output hardware is needed—a printer and a plotter?
5. How easy is it to make changes?
6. Can the software optimise resource allocations?
7. Can the software manage a multi-project scenario?
8. Does the software adopt a hierarchial structure?
9. Is there an integral financial package?
10. Who else with similar requirements uses it?
11. How good is the documentation?
12. Is a benchmark test needed?
13. Will there be continuing support for the hardware and software?

2.3 FINANCIAL CONTROL

2.3.1 Estimating

At several stages during a project, financial estimates are necessary, but the accuracy of the estimate need not be so high in an initial study as in a definitive study. The classification of estimates in table 2.3 is of course subjective.

Estimates for feasibility studies justify some work in their preparation, and a definitive estimate requires and justifies more effort. The definitive estimate will improve in accuracy as a project progresses, and it is unlikely that the extra expense in obtaining more accurate estimates is ever justified.

2.3.2 Controlling Purchased Goods

Most projects will involve the purchase of components or assemblies, and, an appreciation of the legal position (Chapter 4, sections 4.3–4.7) is valuable. A competitive and efficient purchasing system is also essential, and the main requirements are:

(1) to obtain a favourable price;
(2) to ensure ordering at the correct time;
(3) to ensure delivery according to the schedule;
(4) to provide adequate storage and control of the goods received;
(5) to ensure that the quality of bought in goods is up to standard.

It is particularly important to check the quality of goods upon receipt. If a fault is not discovered until an item is installed, then both time and money will have been wasted. Furthermore, a long delay in identifying faults with purchased goods can affect the statutory rights of the buyer (Chapter 4, section 4.4.4). It is of course equally important to check the

Table 2.3 Classification of Estimates

Type of Estimate	Accuracy	Application
Order of magnitude	Within a factor of 10	Limited
Ball park or top of the head	± 25%	Initial study
Comparative (based on previous experience)	± 15%	
Feasibility	± 10%	Feasibility study
Definitive	± 5%	Final planning

quality of goods supplied in-house: the important topic of quality is the subject of the case study in Chapter 9.

By phasing the ordering, not only is the associated work load levelled, but the storage and handling requirements are also levelled, and the cash flow in the project will be helped. Adequate controls are also needed for bought-in items, especially those used throughout a project.

One technique is a relationship based on the ABC or Pareto curve, a form of hyperbola. It is widely believed that for a large range of products, 10 per cent of the items account for 70 per cent of the total cost (A items), while 70 per cent of the items account for 10 per cent of the total cost (C items). The remaining 20 per cent of the items account for 20 per cent of the cost and are designated B items. These three regimes are shown in figure 2.17. Strictest controls should be applied to A items, while C items justify the least control. With C items the policy should be to over-order, since any delay caused by cheap parts would be much more expensive than the extra purchase cost. The divisions between the items are of course somewhat arbitrary, nor is the 70, 20, 10 breakdown an exact, universally applicable

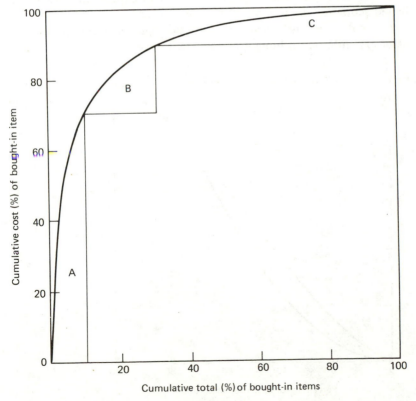

Figure 2.17   ABC curve for bought-in items

law. With the Pareto principle, items are divided into two categories, with 80 per cent of the items assumed to account for 20 per cent of the total cost, and 20 per cent of the items accounting for 80 per cent of the total cost.

2.3.3 Integration of Progress and Financial Monitoring

During the planning of the project, the sequencing of the activities will have been chosen and usually the cost of each activity will have been estimated. From this, the cash flow requirements of the project are determined. When cumulative achievement, cost or man–hours are plotted against time a characteristic S curve is produced; this is shown in figure 2.18.

The *S curve* is a consequence of a slow start to projects, followed by a period of rapid progress, concluding with a reduction in activity to reach the end.

The inevitability of this arises from several reasons:

(1) only a small number of parallel activities can occur at the beginning and end;
(2) the initial activity is often design (which is not capital intensive) with limited scope for the number of people involved;
(3) in the final stage any subsidiary activities will have ended, and the major expenditure on bought in items will have ceased.

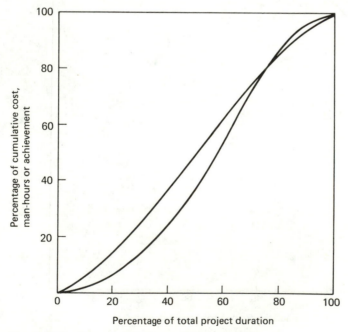

Figure 2.18 Two of the many possible S curves

Paradoxically, it is during the middle stage that delays and setbacks are most likely to occur; though it is the period of most rapid achievement, it is also the period when achievement is least apparent. An appropriate analogy is ploughing a large field—an activity that should reduce the S curve to a straight line. Initially progress is quite visible, since the farmer can see how much has been ploughed, and the rate of change is quite apparent. At the halfway point there are equal areas of ploughed and unploughed land and the rate of change on a percentage basis is minimal. In the final stages the rate of ploughing the unploughed land will again be quite apparent. Thus it is very important to carefully monitor progress during the middle stage of a project, and to maintain the enthusiasm and commitment of the project team. This is a good reason for using the 'milestones' described in chapter 7, section 7.4.

The S curve can be a very useful and sensitive device for the analysis and monitoring of a project; by comparing the progress with the predictions, any discrepancy can be identified at an early stage. A technique described by Cowell (1982) can be used to provide updated estimates of the project cost and completion time (figure 2.19).

The solid line in figure 2.19a shows the original predictions for expenditure and achievement on a project that was budgeted at £600 million, with a duration of 8 time units. In this example, the value of the achievement is originally assumed to be linked with the expenditure. The actual achievement and expenditure are recorded for each time unit and, for this example, 5 times units have elapsed. The expenditure is below the prediction, but the achievement is even lower. Time prediction is made on figure 2.19b; the time datum line represents progress according to the original schedule and thus there is a linear relationship (of unity slope), linking the elapsed time and the programmed time of achievement. Similarly, on figure 2.19c the cost datum line shows a linear relationship (again with unity slope) between the original prediction of expenditure and achievement.

The first step is to produce cost and time prediction lines. The light broken lines show the construction for the predictions made at the 5th time unit. The points on the time prediction line show the elapsed time, compared with the time when that achievement level should have been reached. Similarly on figure 2.19c the points on the cost prediction line compare the actual achievement with the actual expenditure. In each case the broken part of the prediction lines is based on extrapolation; the estimates can be made by eye, or some form of regression analysis.

The next stage is to make predictions of expenditure and achievement projected from the 5th time unit; in this example the heavy broken lines show the prediction for the 7th time unit. The outstanding time is derived by assuming that the outstanding work is performed in accordance with the original prediction, and the outstanding costs are derived by assuming that the cost of the outstanding work is in accordance with the original

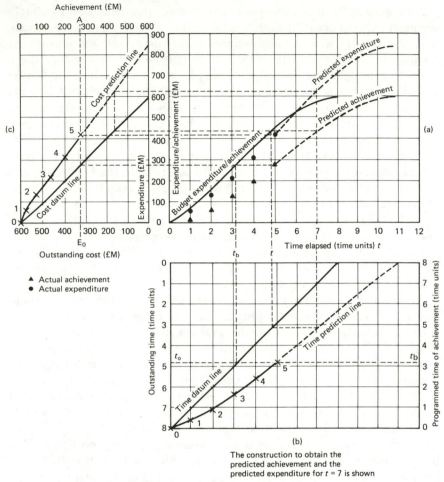

Figure 2.19 *The use of project time and cost monitoring to give better estimates of total project cost and duration [Reprinted by permission of the Council of the Institution of Mechanical Engineers from E. B. Cowell (1982), 'The integration of project time control and project cost control', Proc. I. Mech. E., Vol. 196, No. 28, pp. 313–16]*

budget. By repeating this procedure for other future time units, the lines for predicted expenditure and achievement (broken lines on figure 2.19a) can be added for the remainder of the project.

2.3.4 Financial Contingency Planning

Since projects have a tendency to over-run and over-spend, a pragmatic approach is to look at how sensitive a project is to an over-spend and over-run. It is equally important to know the benefits that might accrue from underspending and early completion—this may lead to a change in

	10% Under-spend	Budgeted spend	10% Over-spend
Completion 10% early	PB_{11}	PB_{21}	PB_{31}
Completion on time	PB_{12}	PB_{22}	PB_{32}
Completion 10% late	PB_{13}	PB_{23}	PB_{33}

Figure 2.20 A payback matrix

the project. One way of examining these effects is to set up a *payback matrix* (figure 2.20)—the use of 10 per cent in this case is of course quite arbitrary.

The use of payback periods implies a timescale of at least several years, and the payback period is calculated using Discounted Cash Flow (DCF) or some other similar technique (see Appendix A). Many projects have a payback period of less than a year and for such projects a DCF analysis is unnecessary. In this case the benefits of finishing early (or late) are an earlier (or later) start to the savings.

One cause of over-spending is currency fluctuations that affect the cost of items or services procured from foreign countries. This problem can be avoided by forward buying. An agreement is made with a banker, who undertakes to suppy the required amount of foreign currency at a rate fixed at the time of the agreement.

Materials are also subject to price fluctuations and there are established Future Commodities Markets for some commodities such as certain metals (copper, tin, silver etc.) and other raw materials such as oil and its derivatives. On a smaller scale, more competitive tenders may be received by a client if he undertakes to supply materials to the contractors.

2.4 CONCLUSIONS

In the first instance, projects require sensible estimates of costs and timing, and from this information a programme can be proposed. Whether the programme is based on a bar chart or a network analysis will depend on several factors, including:

(1) the project size—bar charts are more appropriate for small projects;
(2) the company policy;
(3) the expertise available;
(4) any personal preference.

Since most projects are small, bar charts are very widely used—there is nothing to be gained in using a technique that is more complicated than necessary. Bar charts are also amenable to modifications that show float, interdependence of activities and the resource levels required at each stage of the project. Furthermore, the bar chart has a direct relationship with time, and progress can be compared directly with the schedule.

The other main output from the planning stage is the expenditure as a function of time, indeed this may have led to a re-adjustment of the original schedule in order to optimise the cash flow. The predictions of cumulative expenditure and achievement both usually produce S curves, and these can be used as a yardstick for judging the project performance. When a project diverges from the plan remedial action has to be taken, and the quicker the better. A method of estimating the effect of any delays or overspending on the project duration and final cost has been presented in section 2.3.3.

In the case of projects with a long payback period a Discounted Cash Flow (or similar) analysis is necessary, and it is prudent to know the sensitivity of the payback period to both delays in project completion and over-spending. Of course, some projects may never involve a payback analysis; examples include safety-related projects and amenities such as improved roads. Further examples can be found in the consumer products industries (food, confectionery, toiletries etc.) where marketing considerations may over-ride all considerations of the project financial payback, and even demand that the project be telescoped into the minimum time (regardless of cost). Where such a situation is identified this must be clearly stated, and it may be useful to indicate, in general terms, the premium associated with such devices as extended overtime working and premium payments to suppliers (preferably linked to penalty clauses).

2.5 DISCUSSION QUESTIONS

1. Compare and constrast the advantages and disadvantages of bar charts and network diagrams, for project planning, monitoring and control. List some of the requirements for network analysis software.
2. Discuss the role of estimating in Project Engineering, and identify the strategies for obtaining the most reliable estimates.

3. How are S curves constructed, and what is their role in project monitoring?
4. How can bar charts be adopted to maximise their use in project planning and monitoring?
5. Define the following terms used in network analysis: CPA, PERT, Earliest Start Date, Latest Finish Date, Total Float, Critical Activity and Critical Path.
6. Under what circumstances are dummy activities used in network diagrams? Illustrate the different types of float found in network diagrams, and indicate their significance.

Chapter 3
Management of
Engineering Projects

3.1 INTRODUCTION

This chapter begins with an explanation of the alternative forms of organisation, and their appropriateness for the management of different kinds of project. This is followed by looking at some of the problems arising in project management. The next part of this section examines the skills and attributes required in project managers. The final section looks at aspects of the *project management process*—that is, achieving the project objectives of cost, time and budget through other people. The Bibliography contains details of further reading that covers project management in more detail. The functions of the project manager and project leader were defined in Chapter 1, section 1.1, this chapter assumes that both these roles will be taken by a single project manager; indeed for a small project these and other roles will be taken by the project engineer. A further definition that is needed for this chapter is the *Line Manager*: the person who is responsible for a particular function within a company. The line manager usually has no project responsibilities, and a project manager will request functions to be performed by the line manager's department or section. *Functional Manager* or *Departmental Manager* are synonymous with Line Manager.

3.2 ORGANISATIONAL DESIGN FOR PROJECTS

3.2.1 Background

Figure 3.1 illustrates the continuum of structural forms adopted by organisations, as described by Galbraith (1971). It can be seen that the matrix

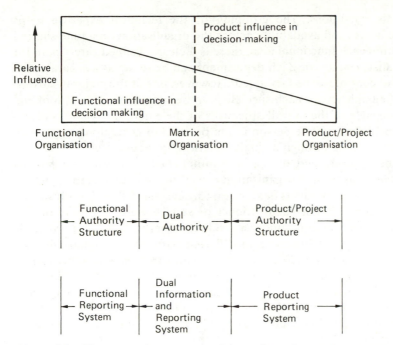

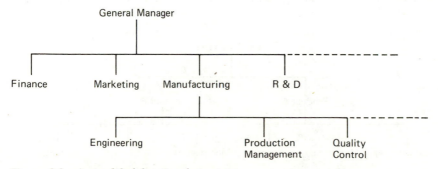

Figure 3.1   The range of organisational forms [Based on Galbraith, 1971]

organisation is mid-way on the continuum between the functional organisation and the product organisation. The *functional organisation* is based on
a subdivision of tasks by function, for example, Contract Administration,
Quality Control, Planning and Costing. Each of the specialist departments
contributes to the common product or service. It is highly suited to an
organisation with a very limited product or service, offered to a limited
number of markets. A simple illustration is given in figure 3.2. This form
of organisational structure is claimed to be economic when applied in

Figure 3.2   A simplified functional structure

appropriate conditions. The grouping of the technical expertise helps maximise its use and availability across the organisation on a time-sharing basis. Within each functional area, there is a clearly defined career path for the specialists employed. Each department can be set up as a separate cost centre, and control can be facilitated. However, one of the major disadvantages is the tendency for each specialism to become rather inward looking, and to lose sight of the overall objectives of the enterprise. There is also a heavy dependence on one person in the position of general manager, with inadequate training through a single function for a successor.

At the opposite end of the continuum, Galbraith shows a *product organisation*. This form of organisation is well suited to an enterprise which produces a range of products or services, for several markets. It is illustrated simply in figure 3.3. For each product or service there is a more or less similar functional structure under a product manager. In this form there is clearly duplication of functions in different parts of the organisation—a feature which leads to additional costs. Against this, each product can be treated as a profit centre, and staff loyalty and commitment to the unit can be achieved; and each product manager is gaining general management experience. The division of the organisation illustrated here is on the basis of products; the division could equally well be based on regional divisions or discrete project teams.

Any project of significance to the organisation, in terms of high financial, political or internal risk, requires a form of organisation structure, that is somewhat different from the hierarchical form that is found in both functional and product organisations which are in a 'steady-state'. This hierarchical form of organisation results from management's pursuit of economies of scale. Rationalisation of production leads to an increased division of labour, which in turn leads to a concentration of resources, people and plant, into larger, more specialised units (that is, *functional departments*) that are integrated by taller *line management* hierarchies, and fatter procedure manuals (Knight, 1977).

Where a hierarchical organisation is faced with the need to innovate (whether in terms of new products, services or systems), managing such a

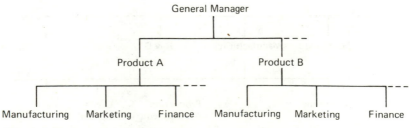

Figure 3.3 A simplified product structure

change may well be unwieldly unless a *project team* or *task force* is set up, in which the various interested parties are represented. In a similar way, where the needs of a project client are such that several areas of expertise within the supplier organisation or contractor are needed to fulfil the contract, that client will expect to see a clear single point of responsibility (the project manager or contract manager) within the supplier organisation for carrying out the task.

Effectiveness in managing projects is normally the result of a project organisation which encourages high levels of communication, flexibility and responsiveness; rapid decision-making; the resolution of unstructured problems; and effective planning and control in conditions of uncertainty. The matrix form of organisational structure (figure 3.4) is often advocated for such tasks. The first application of the matrix structure is accredited to the United States aerospace industry in the late 1950s, as a result of government requirements for single point responsibility among and within the contracting organisations. Nearly thirty years of experience has been gained of their operation, and many books on the subject of project management describe matrix organisations in detail. The remainder of this section starts by looking in some detail at the various forms of matrix organisational structure, before going on to some practical examples.

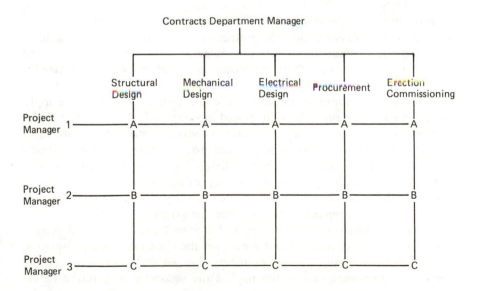

A — Staff involved in Project 1
B — Staff involved in Project 2
C — Staff involved in Project 3

Figure 3.4   A simplified matrix structure

3.2.2 The Different Types of Matrix Organisation

Knight (1977) identifies three types of matrix organisation:

(1) the co-ordination model;
(2) the overlay model; and
(3) the secondment model.

In the *co-ordination model* all the staff remain in the department in which they are normally based, but organisational arrangements are introduced, to enable cross-departmental collaboration for the purposes of the project. Where a project team is formed, it would normally have representatives from each of the departments or sections involved in the project, with these representatives reporting back to the team on progress within the particular departments. The project leader in such situations appears often to be in a co-ordinating role rather than managing. Knight points out that a weakness in this model is that those being co-ordinated always have the right of direct access through their line management to the higher authorities who are setting or sanctioning the tasks to be co-ordinated.

Staff become members of two organisational groupings in the *overlay model.* This is illustrated in figure 3.4 where the members of functional departments within the organisation are engaged on a number of projects. Responsibility for the professionalism of the specialists normally rests with the functional manager, while the project manager is responsible for achieving the project completion within the budget. Each staff member may find himself working on more than one project at any given time, and hence for more than one project manager. This type of structure is not uncommon in research and development operations and in engineering design. The application to computer projects is illustrated in Chapter 7, section 7.1. A major source of difficulty in this arrangement is often the maintenance of agreement between the numerous managers involved. The balancing of an individual's workload and the measurement of individual performance can also be a problem where staff are involved in a range of projects with several bosses. The individuals concerned will often find difficulty in determining priorities where demands are being made by several managers.

The *secondment model* involves staff moving from a functional department to the project team or task force, for the time taken for completion of the element of the project requiring the particular expertise of the secondee. On completion of the project the secondee may return to his original functional department. This is quite a common form where the projects are of considerable duration, such as the construction of a power station or a petrochemical complex. In such cases, the secondee may never return to the functional department, because of a shift in personal career aims, or in some cases the lack of work in the original department. This can cause personal problems, where the individual feels out of touch with those more senior, who make decisions about resource allocation and promotion. As the secondment model is used for large-scale projects, a hierarchical structure will often be necessary within the specialist functions of the project team. In organisations which use the secondment model for managing projects, the project manager is normally in a powerful position within the organisation, and is influential in the choice of resources to be assigned to his particular project.

3.2.3   The Matrix Structure in Practice

The use of a matrix structure for the organisation of projects is not without its problems. Handy (1981) describes the outcome of research at Corning Glass into the management of project teams. It was reported that project teams are a waste of resources when a task can be handled effectively through the existing structure. The task must be a clearly defined project so that participants have a clear goal to aim at. Unless others in the organisation see the project as worthwhile, it will not be adequately resourced. The project team will need protecting from the surrounding bureaucratic organisation. Not only must the team members be professionally competent, but they must also be effective team members. The team also has to be capable of coping with its changing membership. External pressures from colleagues and superiors in the functional organisation have to be resisted, and the team members have to accept responsibility towards the team and its objectives.

The overlay matrix in particular is trying to optimise two conflicting benefits as pointed out by Child (1984). On the one hand it aims to retain the economic operation and technical capability associated with functional organisation. On the other it attempts to co-ordinate those resources in a way that applies them effectively to different organisational activities. However, most writers about organisation structure seem to share the view that a matrix structure is only an appropriate form in extreme circumstances. These are listed by Davis and Lawrence (1977) as:

(1) when two or more sectors are critical for the organisation's success (functions, products, services, markets, areas);
(2) when there is a need to carry out uncertain, complex and interdependent tasks;
(3) when there is a need to secure economies in the use of scarce human resources, such that their use is shared and flexible.

The use of matrix and hierarchical structures can be usefully illustrated by two very different examples. The first is the organisation structure of a major engineering contractor; two organisation charts representing the AJ Corporation are illustrated in figures 3.5 and 3.6.

Within the AJ Corporation there are three main companies shown, including the International Corporation which includes the British company. This company in turn is divided into separate companies on a product split. Other subsidiaries of the AJ International Corporation are located in various countries throughout the world. However, the split is not based on a particular market configuration, as much of the corporation's business is not located within these countries, and each is not assigned a specific and exclusive territory. The various national operations will combine resources to compete for projects in response to particular project needs; for example, one company may be better placed to arrange project funding for a major petrochemical complex, whereas another may have a particular design expertise available. Within AJ Energy Ltd, as illustrated in figure 3.6, there is a functional split. AJ Energy manages the design and construction of process plants for the petrochemical industries, ranging in value from £5million to £500+million. Success as a company depends upon their ability to project-manage a range of contracts. For each project a project manager is appointed, and a team is formed from members of the various

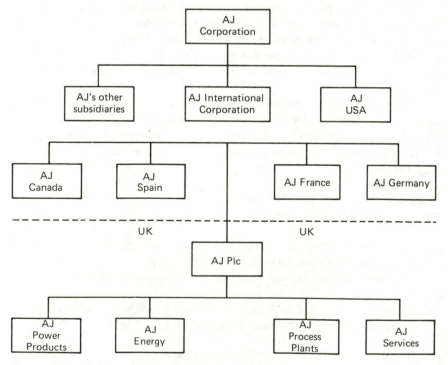

Figure 3.5   AJ Corporation organisation structure

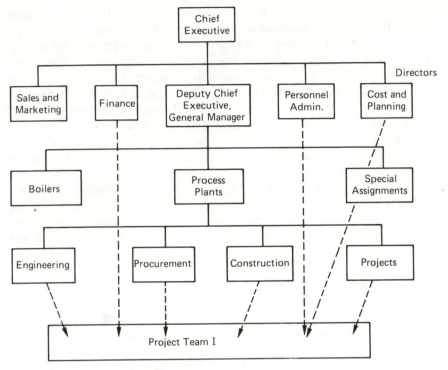

Figure 3.6   AJ Energy Plc organisation structure

specialist departments needed to meet the project objectives (figure 3.6). In some cases the team will be located together, as is the case at the construction stage of the project.

This is in line with the secondment model outlined earlier. In other cases the individual specialists will be involved in several projects consecutively, and the operation will resemble the overlay model.

In practice, a mix of organisation structures can exist within one organisation, even though much of the company is involved in providing project services to a number of clients at any one time. A closer look at any one project would probably show that the structure changed during the life of the project. The factors determining the type of structure that is appropriate at any stage include the project size, the nature of the key activities, the decision-making requirements, and the client's needs in relation to handover and commissioning. The preparation and tendering stages of the project cycle normally require flexibility, high levels of communication between the various disciplines and a high level of professional competence. The mid-phase in the project cycle can more readily be programmed and subdivided into clearly delineated areas of responsibility. The optimum structure for each phase may well vary, but the choice of an organisational form for the project will also have to take into consideration other factors, such as the ability of those involved to cope with such changes.

The AJ project manager has to concern himself with a wider project organisation than is reflected in the AJ organisation chart. The wider organisation of the project is illustrated in figure 3.7. This shows the relationship between the project client, AJ's project organisation, suppliers and sub-contractors. At the very least, each of these parties has to understand to whom they relate in AJ's structure, and to seek to be clear about the authority vested in that individual. The client organisation, as well as major sub-contractors, will have their own project organisation often headed by a project manager.

The second example is chosen to illustrate an organisation for project work that is based on a food manufacturing company. The structure of the confectionery division is shown in figure 3.8. While this is one of a number of divisions within the company, it has its own engineering department that is responsible to the manufacturing director for the development and execution of any engineering projects within the company. The projects range from a multitude of small-scale modifications to manufacturing plant, to major projects involving the replacement of complete manufacturing units within very tight time periods. The engineering department includes a number of specialisms as shown. In addition to engineers, it employs tradesmen covering each of the main trades involved in the type of projects undertaken by the company. The range of trades is rather less than would often be the case, because the company is non-unionised, and has had some success in negotiating multi-skilling. The planning section within engineering is responsible for scheduling the work. At any one time it is likely that a wide variety of projects will be under way concurrently. Each project will require the services of more than one discipline. Each project is assigned

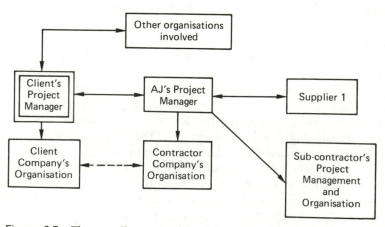

Figure 3.7 The overall organisation of the project [Based on Harrison, 1985]

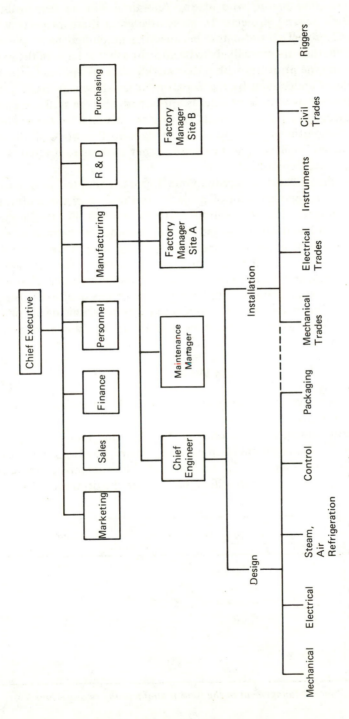

Figure 3.8   A food manufacturing organisation

to a *project co-ordinator*, who at any one time will be responsible for co-ordinating several projects. In some instances these projects will be allocated to project co-ordinators because the individual has a particular expertise, but they are more likely to be allocated on the basis of the existing workload and the priorities within that workload. This applies to the many small projects undertaken by the department, as shown in figure 3.9. For large-scale projects a project leader is appointed, and the staff are assigned to the project on a full-time basis. During the installation the tradesmen within the department are appointed to oversee the work of the many sub-contractors involved, and to oversee other tradesmen drafted in from the maintenance department.

Again, this illustrates more than one approach adopted within a single company for the management of projects. In the case of the multitude of small-scale projects, a co-ordination matrix is adopted. For the large-scale project a secondment matrix is used. Clearly included in the factors leading to the secondment approach for the large-scale project are the relative complexity of the task, the tight time pressure, the potential costs of a project over-run, and the need for meticulous preplanning that presupposes clearly defined responsibilities and priorities. For the small-scale projects, the co-ordination model offers a cost-effective way of organising the projects, where the time pressures are much less and the technical expertise needed is rather limited.

3.2.4 Making Matrix Methods Work

This section started by looking at the matrix structure as a means for managing projects (3.2.1), it then identified three types of matrix structure (co-ordination, overlay and secondment) (3.2.2), and finally the approach adopted for the management of engineering projects by two different types of company was described (3.2.3). Some of the problems in operating a matrix structure have been identified, but despite the drawbacks associated

Chief Engineer	Structural	Electrical	Mechanical	Steam/ Air Refrigeration	Control	Packaging
Project co-ordinator 1	×	×	×	×	×	
2		×	×		×	
3	×	×	×	×	×	×
4					×	×

Figure 3.9 Project Management in the food manufacturing organisation

with matrix management it is widely used in one form or another. This section ends by drawing together some of the key factors to be considered when structuring the project organisation. However, it must be recognised that each project will have its own peculiarities, and each will demand a unique solution.

The main points to be considered in the decision about the project management organisation are listed in table 3.1.

While organisational structures, represented on paper in the form of organigrams, can be helpful in describing the form of that organisation, they will never be able fully to reflect the actual operation of the organisation.

Table 3.1 Summary of Considerations when Designing a Project Organisation

(a)	Matrix structures may be appropriate where there is a need for communication across traditional disciplinary boundaries at various levels in the hierarchy; a high level of flexibility is demanded of the staff; the organisation is required to respond rapidly to meet the changing situations, and to resolve complex problems
(b)	The project co-ordination model may be useful where technological resources are scarce and have to be shared between a number of projects, and where the development and retention of an expertise in the various disciplines is essential for the organisation
(c)	In the project co-ordination matrix the project leader is normally in a co-ordinating role, and in a weaker position within the organisation than the functional manager, and difficulties may be experienced in establishing priorities between the competing projects. The project co-ordinator will not be able to resolve these conflicts, but depend instead on higher authority to enforce the decisions
(d)	The secondment model may be appropriate where projects extend over a long period, are complex and when uncertainty is high
(e)	A major drawback of the secondment model is seen to be the separation of the specialists from the specialist team. This can result in the slower development of expertise within the specialism, as well as uncertainties for individuals concerning their personal career development and job security on termination of the project. Some members of the team may well have been recruited for the specific project, and be on a fixed-term employment contract for the duration of the project
(f)	In the secondment model the project manager normally has considerable responsibility, and is in a powerful position organisationally compared with the functional managers
(g)	The overlay model seems particularly appropriate for situations where the projects are of a short duration, are complex, and where several concurrent projects depend on scarce specialist resources
(h)	In practice the overlay model may not be sustainable in the long term. The organisation is likely to move in the direction of either the co-ordination matrix or the secondment form. In the case of the secondment form, the project managers will strive to maintain a constant team, and to ensure sufficient project work to justify its maintenance; the organisation will then become more like the product/project form suggested by Galbraith

In addition to the formal structure with its formal reporting relationships, there are informal systems operating within the system. Dalton (1959) defines formal or official as 'that which is planned and agreed upon' and informal or unofficial as 'the spontaneous and flexible ties among members, guided by feelings and personal interests indispensable for the operation of the formal, but too fluid to be entirely contained within it'. It is common place for managers to bypass the formal system of communication represented in the organisation chart, and to build their own network of informal contacts. These informal networks are often quicker than the recognised channels, and the information gleaned will often contain a sense of the prevailing climate relating to the issues in question. These informal channels will provide social satisfactions for those engaged in the encounters but, perhaps more importantly, they will be used as a way of enhancing a personal position. Without these informal systems, organisations would not function. These systems are particularly important in the management of projects, because no organisation chart can possibly define adequately the roles to be performed to meet the project objectives. Staff engaged on projects have to be flexible in their approach to work, otherwise the tasks required over the life of the project will not be accomplished.

To conclude, despite all the problems associated with matrix structures, they are in many instances essential in the management of projects. Stucken-bruck (1981) suggests a number of steps that are required to ensure that matrix structures work:

(1) top management must give real and immediate support to the matrix, including a clear brief which states unambiguously the objectives for the project, and the responsibilities and authority of the project manager;
(2) the brief should spell out, as far as possible, the project manager's relationship to the functional managers involved;
(3) functional managers must modify their thinking and managerial behaviour to ensure that the matrix works;
(4) the project manager must exercise negotiating skills to ensure the compliance of the functional managers to the project plan;
(5) project personnel must be able to adapt to a situation of two bosses.

3.3   THE PROJECT MANAGEMENT TASK

3.3.1   Introduction

In a survey of over 50 project managers and more senior executives, Birchall and Newcombe (1985) identified a range of key problems facing those responsible for managing projects. This section will examine those areas in turn. Their order of occurrence reflects the project life cycle (figure 1.1), and they are: establishing the project brief, defining the role of the project manager, the management of design, and the management of design implementation.

3.3.2   Establishing the Project Brief

The first task in managing a project is that of establishing the brief for the project. In the case of projects that are large in relation to the total company assets, or where the potential for disruption of on-going activities is high, the decision to proceed will be taken at Board Level. The objectives for the project will relate to a corporate plan, either explicitly or implicitly. In the case of less significant projects, the authority to proceed will be delegated to a lower management level, but it will still be subject to a review procedure and specified criteria, such as a defined payback period (Appendix A). Despite this decision-making process, the full objectives for the project will often not be made generally explicit, with perhaps the concealment of sensitive commercial information, or the deep-rooted personal objectives of the decision-makers. The project manager is fortunate if he is a party to this decision-making process, in that he is then more likely to appreciate fully the objectives of the project. Such a high level of involvement of project managers at this stage is positively resisted in some organisations, in order to avoid the project manager becoming too committed to objectives which are unattainable, and in consequence losing a degree of objectivity at the later stages in the project.

Whether involved from the earliest stages of the project or not introduced until a decision to proceed, the project manager must establish a clear brief. This *project brief* should not only be concerned with the aims of the project in terms of the performance specification, time and cost limits, but should also establish the role of the project manager. Often the initiators of the project need assistance in clarifying the objectives and forming a realistic assessment of the viability of the project. This is particularly likely to be the case where the decision-takers have little or no prior experience; for instance, where a company seeks to build a new factory for the first time, and does not employ staff with expertise in managing building projects.

3.3.3   Defining the Role of the Project Manager

The project manager must establish a clear definition of his own role. Where the person appointed is an employee of the organisation sponsoring the project this is essential, because the individual is seeking to build a career in the company, and he cannot afford to be held responsible for mistakes over which he had no authority or influence. As an appointed consultant project manager, the individual and his employer will be taking on duties which, if performed negligently, could lead to redress in a court of law (Chapter 5, section 5.5). It is therefore vitally important that the role is clearly defined and, where relevant, adequate professional liability cover should be obtained from an insurance company.

At this early stage in the project, the project manager or engineer is likely to have to work with a range of professional specialists. In the case of a new product development this might include research and development, marketing, production engineering, specialist materials and equipment suppliers, and a host of others. In the case of property development, the specialists might again include marketing; but architects, structural engineers, quantity surveyors, services engineers and planners are almost certain to be required. Each of the specialists is likely to be immersed in his own specialist jargon, and his own particular field of expertise. The project manager has to ensure that the project benefits to the full from the range of expert advice available. To do this, the project manager must be able to establish credibility and respect. The project manager is the person who will ultimately have to balance the sometimes conflicting objectives of performance, time and cost, and ensure that no one element of the project gets unwarranted emphasis—for example, the car does not have a superb gearbox at the expense of an inferior engine.

Given the high level of uncertainty in the market place for many products and services, the prediction of demand is problematic. Consequently project managers will often have difficulty in establishing the project's viability. Persuading decision-makers to invest at times of high interest rates and business uncertainty is also difficult, even though on the surface it appears likely that prices charged by suppliers will be low when the volume demand is low. The sponsor for the project, often referred to as the *project champion*, needs not only to produce a soundly argued case on technical grounds, but also to appreciate how decisions are arrived at, and the political processes that are involved. If the project champion has a good all-round understanding, the prospects of obtaining the sanction to proceed can be improved, as well as the prospects for the project overall being seen as successful. Understanding the total picture, concerning project objectives, is clearly essential when attempting to push through sanction for a project.

3.3.4   The Management of Design

Following approval to proceed, the project manager has the task of managing the design phase. Decisions made during this phase will have a major impact on the final cost of the project, the operational costs and the eventual costs of dismantling or demolition. The project manager has to have an understanding of the implications of decisions made at this stage, on not only the overall project costs and life-cycle costs, but also on the time to completion.

The specialisms required for design will relate to the type of project being undertaken. It is important, however, to identify the specialisms required, to assess the capability within the organisation, and appoint suitable consultants where necessary. In selecting consultants and other members of the design team, the project manager should use a range of criteria. Important among these is professional competence, but clearly fitting into the design team, and an ability and willingness to work with the other team members, are also very important.

Designers and the technical problem-solving process are not easy to manage. It is often difficult to put realistic time estimates on to activities, and as a result the work programmes produced need careful monitoring, and corrective action must be taken where deviations look certain to prolong the activity. The project manager should identify those areas where the project is most at risk, concentrate the monitoring activities in these areas and have contingency plans for coping with over-runs. Any work programmes should have milestone dates for achieving the key activities, and project management must keep the pressure on the team to achieve these. The use of milestones and other techniques such as brainstorming are discussed further in Chapter 7, section 7.5. Many designers resent close monitoring of their work, therefore the project manager needs to establish the ground-rules for team membership at the outset, and then demonstrate throughout that the ground-rules are to be adhered to. Many of the team members might also be employed on other projects where their expertise and time is required. These other projects are unlikely to be under the control of the same project manager. In consequence, each project manager is likely to be applying pressure on individuals, and seeking the top personal priority for his own project. This will cause tensions at times which will have to be resolved. The project manager has to evolve strategies for resolving these tensions without interfering with the work programme of his own project. This is discussed in Chapter 7, section 7.4, in connection with computer projects.

The effective project manager at the design phase is not only concerned with the internal workings of the design team, but also with ensuring that its work can proceed with the minimum of interference. The project manager is concerned with anticipating barriers to progress, and to their elimination or establishing a working compromise. These barriers to progress may well be within the organisation—for example, other managers who do not wish to see the project succeed, and attempt to starve it of resources—but it is more likely to be an external threat. A range of regulatory bodies, such as Local Government Planning Authorities and the Health and Safety Executive, may have to be dealt with. Various other interest groups may seek to shape the outcome. In the case of large-scale projects, such as the United

Kingdom Central Electricity Generating Board's first pressurised water reactor at Sizewell, the machinery for dealing with the issues raised by these interest groups delayed progress considerably.

In areas where the technology is changing rapidly, a decision has to be taken to adopt a particular solution, even though possibly a more appropriate means will shortly be available. Designers generally take great professional pride in being at the forefront of new technology, and have a strong desire to incorporate the latest gadget into their design. The project manager has a responsibility for achieving an acceptable solution to the brief, which is unlikely to stipulate the incorporation of the latest technology, and in reality this may be seen as less than desirable if the reliability is uncertain as a result. Consequently, the project manager has to monitor not only progress, but also the basis of design solutions. Also of importance to the overall progress is the clear definition of the design freeze. Many large-scale projects undertaken in the United Kingdom during the 1970s suffered time and budget over-runs which were partly attributable to late design changes. Clients now usually prefer to freeze the design at a planned stage, proceed with assembly/construction and make minor modifications on completion, if this is required by changing legislation or a justifiable technological advance.

3.3.5   The Management of Design Implementation

At the execution stage, involving procurement, erection or construction and commissioning, the nature of the problems to be overcome changes. Materials and equipment supply problems can cause considerable delay unless anticipated, and provisional orders may need to be placed at an early stage. Difficulties were reported by respondents to the survey (Birchall and Newcombe, 1985), even in obtaining the necessary information from potential suppliers regarding technical details, let alone a commitment to delivery dates and performance specifications.

On larger projects, much of the work is packaged into parcels and is undertaken by sub-contractor organisations: this leads to problems of co-ordination. Care is needed in formulating these sub-contracts, so as to achieve consistency with both the main contract between the client, main contractor and sub-contractors, but also between the sub-contractors. Particular areas of concern are on the interface between the various contracts, where time delays in one job interfere with the work of subsequent contractors. Problems are also likely where various contractors have to share common facilities such as cranage.

3.3.6   Attributes of the Project Manager

Project success depends on selecting the right person as the project manager. The more significant the project is to the organisation, the more important is the selection of the project manager. So what are the qualities needed for effective performance in this key role?

As part of the survey referred to earlier (Birchall and Newcombe, 1985), information was collected about the key attributes and skills needed for effective project management. The results were classified into technical and commercial expertise, personal attributes and skills.

The survey revealed considerable emphasis being placed upon the need for broad-based technical expertise. An understanding and experience of all the phases of the project enable the manager to identify problems at an early stage. While not expected to have performed in all the specialist roles himself, the project manager must have an appreciation of all the specialist areas required by the project.

Specific technical skills that were identified included an understanding of computers in project management, planning techniques, and methods for monitoring and control (these techniques were described in Chapter 2). Project managers were required to have strong commercial expertise. The job of project manager not only involves harnessing and managing resources committed to the project, but also managing within the wider political and organisational context. Understanding the way in which the client 'ticks' and understanding the constraints under which the client is operating are essential.

Specific areas of commercial expertise include contracts and contract law (Chapter 5), procurement methods and processes, estimating and cost control (Chapter 2). A general financial awareness, as well as an understanding of project financing, were also seen as important.

Among the personal attributes that were identified, leadership abilities were listed by more respondents than any other single area in the study. Related to leadership qualities were several more specific personal qualities—that is, high personal motivation, stable personality, integrity, dedication, commitment and determination. An ability to turn an apparent failure into a success demands particular leadership qualities, in addition to a high level of resilience. An ability to sense political difficulties, and to demonstrate judgement on when and when not to intervene, is also essential.

Project managers, to be successful, need a breadth of vision, being able to understand the complexities of the project and the inter-relationship between the many elements. The project manager needs continuously to review the many relevant internal and external factors impacting on the project, to identify the significant areas for likely adverse effect on project achievement, and to be prepared to take corrective action. It is also essential that the project manager does not get too engrossed in any one aspect of the project. His job is not that of resolving specific technical difficulties, but rather of managing the specialist in the disciplines concerned and ensuring that a workable solution is found at an acceptable cost and within time.

The survey respondents felt that the project manager must be orderly and well organised, and capable of selecting important information easily from a mass of data. He must be able continuously to audit the work of others, be quick to spot danger signals from hints that problems are occurring, and be capable of learning from both personal experience and that of others.

Turning now to the more specific skills needed by project managers, an essential requirement is an ability to communicate effectively with people at all levels in the project. This includes a high level of negotiating skill, since this forms much of the project managers work. Some negotiations will be conducted in formal meetings, for instance, with the project sponsor, suppliers and sub-contractors. Other negotiations will be undertaken in less formal settings, and will involve persuading colleagues to provide resources, encouraging subordinates to take on additional duties and dealing through the informal networks described earlier. While the outcome of formal meetings will be carefully recorded, the results of informal discussion may remain unrecorded, other than perhaps in a diary note. As so much of the project manager's formal work is done through meetings, an ability to plan, organise, control and record meetings is vital. Effectiveness in this area of work is also based on the application of political awareness and skills.

Leadership qualities were identified earlier as a key attribute; the project manager has to be able to weld staff into a team. Important specific skills include personnel selection, briefing and performance appraisal. To be effective in leadership the project manager has to have a feel for the personal needs of key team members, as well as the overall mood of the team. He must possess and apply a repertoire of responses to situations, and be expert at deciding which type of response best fits the particular circumstances. The formation and management of project teams is described in the next section.

3.4 PROJECT TEAMS AND THEIR LEADERSHIP

3.4.1 The Purpose of Project Teams

Each of the previous sections has emphasised the importance of the team performance in determining the overall project success. Also, the survey identified team leadership ability as probably the most important quality needed from the project manager. This final section looks in some detail at both of these aspects. Clearly they are inter-related and therefore there will be some overlap in the discussion.

A project manager will be concerned with many formal groups during the course of the entire project. The project clients or sponsors will require regular formal meetings, to ensure that the project is progressing towards their goals, and they will want to make decisions for corrective action when necessary. There will be regular meetings with those contributing to design, supply, erection, inspection and commissioning. The project manager may see himself as a member of a number of teams, all of which have as their aim the completion of the project within the specified time, cost and performance limits. However, members of each of these teams may well have additional objectives. For example, contractors and suppliers are seeking to make a profit for their own employers, whereas the client's representatives are aiming to minimise costs for the client.

Each of these formal groups may be seen as a team, in that they share a common purpose (that of achieving project objectives), each has a defined membership criterion in that each member has a legitimate reason for membership, and each has predetermined hierarchies established through the formal constitution. These are the conditions listed by Handy (1981) as a definition that a group exists. However, the project manager is also dealing with less clearly defined teams within the project organisation. He might well describe the total organisation as the 'team'. He is certain to have considerable contact with the senior members of the organisation, both at formal meetings and informally. This is likely to be the most important of the internal teams with which he deals. The issues dealt with by this team will change through the life of the project, as might the membership. The early emphasis may well be on issues of technical viability and budget costs, whereas as the project progresses, the emphasis will shift to supply problems, cost control and completion dates.

In general terms, groups are set up within organisations for a variety of purposes (Handy, 1981):

(1) for the distribution of work among the various members, and for monitoring and controlling progress in its completion;
(2) for problem-solving and decision-taking;
(3) to pass on the decisions taken in other groups or information relevant to the project, to those who need to know;
(4) for information and idea collection;
(5) for testing and ratifying decisions;
(6) for co-ordination and liaison between the many arms of the project;
(7) for increased commitment and involvement of members in the planning and decision-making processes;
(8) to enable conflicts between the various parties and/or levels to be negotiated and resolved;
(9) for inquest or enquiry into the past.

The project manager is likely to be involved in groups, each concerned with one or more of the above purposes.

Individual members of groups have their own particular needs. Membership of groups serves to satisfy social or affiliation needs. Groups also offer the opportunity to establish a personal identity and status. Other members can provide assistance in solving problems, and provide support in difficult times—anxieties about aspects of the project can be shared. Also, a personal sense of achievement can be gained from the collective achievement of the team at the end of the project.

The activities of groups are normally divided into two elements:

(1) The *TASK*—in the case of project teams (that is, the organisation's brief for the group) in relation to the overall objectives for the project.
(2) The *PROCESS*—the way in which the team functions, in terms of such aspects as the decision-making processes, the allocation of responsibilities in order to achieve the end result, the carrying out of activities needed to maintain the personal commitment of members, and finally the way in which members relate to each other.

Teams or groups will vary in their effectiveness in achieving both organisational purposes, and in meeting the personal needs of members. Effectiveness will depend upon a number of factors:

(1) Task factors—including the perceived relevance of the task by group members; the nature of the task in relation to the size of the group (for example, negotiation may prove impractical in a large group); the clarity of the task; the urgency with which the task has to be accomplished.
(2) Resource factors—including the ability of individual members; the range of abilities in the group; the overall size of the group needed to encompass all the skills necessary in relation to the task to be achieved.
(3) The environment—to include the standing of the group in relation to the rest of the organisation; the standard way of operating such groups within the organisation, and hence the expectations; the standing of the leadership within the organisation; the physical location and layout for meetings.
(4) The leadership style—the processes and procedures adopted for task achievement and the motivation of group members.

3.4.2 The Assessment and Formation of Project Teams

How then can the effectiveness of a team be measured? Quality of output, quantity of work, speed of delivery, cost, man–hours consumed, profit? The criteria to be employed will influence the choice of style of operation or 'process'. If members see the style as inappropriate, this will impact upon their personal motivation, and in turn their contribution to the task. Some criteria against which the effectiveness of a project team can be assessed are listed in table 3.2.

The project manager will also need to establish a number of formal groups within his project organisation, as well as to create a group identity

Table 3.2 Characteristics Which Can Be Used for Assessing the Effectiveness of a Group (from Carnall, 1984)

(a)	Members trust each other or express their lack of trust
(b)	The group has clear and specific goals, jointly determined by the members and the leader
(c)	Most members feel a sense of inclusion
(d)	Participants talk directly to one another about what they are experiencing
(e)	Group members share leadership functions
(f)	Members risk disclosing threatening material
(g)	Members recognise, discuss and often resolve conflict
(h)	Members feel that constructive change is possible
(i)	The group has high cohesion
(j)	Members identify with one another
(k)	Group norms are developed co-operatively by the members and the leader
(l)	Group members use out-of-group time to work on problems raised in the group

for all those associated with the project. The points worth bearing in mind in setting up and running these groups are:

(1) teams should have clearly stated objectives, and the brief should also define the team's relationship to other parties in the project;
(2) the style of chairmanship should be appropriate for the type of task in hand;
(3) the size of the group generally should be the minimum required to achieve the task;
(4) teams are provided with the necessary skills to achieve the task;
(5) the task is not achieved in the short term at the cost of the personal needs of team members;
(6) sub-groups not chaired by the project manager are given the necessary status and support, and their output is acted upon;
(7) the life of groups is not extended beyond their usefulness in order to avoid a sense of time-wasting;
(8) members are adequately recognised for their individual contribution to group effectiveness in order to maintain the motivation of team members.

3.4.3 Project Team Leadership

Turning now to the question of leadership, the role of the project manager as a leader has been reported earlier to be an important determinant of the project success. The subsequent discussion of teams, in relation to the project, has served to emphasise the importance of team leadership. But what is expected from the leadership role of the project manager?

The leadership role is categorised by Handy (1981) as:

(1) mobiliser and activator of groups;

(2) ambassador;
(3) model.

It has already been shown that the project manager will execute his function through numerous groups—some of which he will chair, others will be under his direction, and some of which he will be a member. These teams will have different types of purpose and membership. The styles of the leadership required to ensure the optimum use of the resources involved is also likely to vary. In addition to the task and the type of subordinates involved having a bearing on the most appropriate style of leadership, both the preferred style of the leader and the environment in which the team is operating, will also affect the leadership style.

Leadership styles range from the authoritarian to the democratic. The autocratic manager makes the decisions and announces them to subordinates, whereas the democratic leader permits subordinates to function between defined limits, and involves them in the decision-making process. It is now widely recognised that there is no one 'best' style of leadership. The autocratic style, with close control exercised by the leader, is inappropriate where the subordinates have a preference for more self-control, and the task is complex and ill-defined. This might well apply at the design stage of the project, when the leader will have to achieve a balance between the needs of the professionals for self-control and the needs of the project with its time and cost constraints. Later in the project, when the task is more clearly defined, more structured ways of working are possible. When considerable numbers are involved, of both direct employees and sub-contractors as well as suppliers, the style of project leadership required in much of the work will be more authoritarian.

If the leader is powerful, he will often impose on the subordinates his preferred style of operation. In doing so he may well lose a degree of group effectiveness, by ignoring to some degree the personal needs of his subordinates. However, they may be willing to sacrifice satisfaction of some of these needs if they see other long-term benefits as likely to accrue. Being associated with a project success may well be seen as sufficient benefit for them to remain involved. However, the project manager may have difficulty in operating such a mis-match if the project is not going well, and is of low profile within the organisation. Also, a project manager may have difficulty in adjusting his style to the requirements of the task and the team. Once patterns of working have been established for the project, it may be difficult to explain a change in style.

Factors affecting the leader's preferred style include his personal value system (how he sees the leader and who should make decisions); his confidence and trust in his subordinates; his habitual style; how significant he sees his own contribution to be in terms of his personal aims, and the needs of the project; his need for certainty of outcome in any situation; the degree of stress and tension he is experiencing; and the stage in his career (older people tending to be more autocratic).

Turning now to the team members, their preferences for a particular style of leadership will depend upon factors such as their own perceptions of their competence and professionalism; their personal psychological contract with the team and its leader; their interest in the task and perceptions of its importance in achieving the overall project objectives; their own needs for clarity and structure in any situation; their previous experiences separately and together; the culture to which they are accustomed.

The nature of the task will also influence the appropriateness of the style. In situations requiring creativity and the solving of unstructured problems, the democratic style seems most suitable. Where the timescale is tight, a more autocratic style is often essential if deadlines are to be met—group decision-making takes time. If the task is complex technically or conceptually, a democratic style works well, whereas when organisational complexity exists, a more structured approach is essential. Tight control is also needed where mistakes cannot be tolerated.

In addition to considering these factors when establishing the working style for the project, the project manager must take into consideration the norms and expectations of the various other organisations involved in the project. Not only is the project manager part of a wider organisation, with its own requirements of its project managers, but the client will have views on what is an effective approach. As the client or sponsor is meeting the bill, he will be concerned to see competence (as he defines it himself) in the project management.

The project manager has a difficult task if he sets out deliberately to modify his own behaviour in his dealings with all the teams and committees with which he will be involved throughout the project. However, the skillful project manager will be doing this without giving much through to factors influencing his method of working. Rather like driving a car, many of the manoeuvres that seemed so difficult when being taught how to drive come as second nature to the experienced. Taking the analogy further, many bad driving habits remain uncorrected, and periodically a re-examination of practice can lead to a subsequent correction and improvement.

The second major aspect of the leader's role is that of ambassador. If the project manager creates a feeling of confidence in the performance of the team, both in the employing organisation and with the client, it is much more likely that a climate will be created which is conducive to effective team performance. The confidence of senior management is important, particularly when questions of resourcing are raised. A view that the project manager can be trusted only to demand more resources with good reason is more likely to result in their allocation. The awareness among group members that their project manager is well respected will also give them confidence and act as a motivator. Perhaps of even greater importance is the project manager's dealings with the client, and whether or not the employer trusts the project manager in his dealings with the client. Can he 'think on his feet' in tricky situations? Is he well prepared, or does he get 'caught out' by the client's representative? Does he 'put his foot in it' in front of the client?

The last part of the leader's task is to act as a model. Others aspiring to the project management role will look around for successful project managers on whom to model their own behaviour. For the employing organisation, this is an important part of the project manager's job, since the future success of the business will in part be determined by the later generations of project managers.

This section has examined those factors which influence the success of project teams. A key element is the performance of the leader, and hence the concentration on leadership in the last part of the section. Three aspects of the leader's task were explored—namely mobiliser and activator of the group, ambassador, and finally model. While most attention was paid to the first aspect, it was clear that the other two were also of considerable importance, and that in judging overall performance as a leader, none of the aspects could be treated in isolation.

3.5 SUMMARY AND CONCLUSIONS

This chapter has looked at some of the problems of managing projects from a behavioural viewpoint. This led to the concentration of attention on both the structuring of the project organisation and the management of the people within it. The focus in the study of the management of people has been upon the project manager, and his choice of style in managing the many teams involved in any project.

Early in the chapter a distinction was made between the management of a project undertaken within an organisation, as part of its capital expenditure programme, and a project organisation where the sole purpose is to carry out projects on behalf of others—the client/contractor relationship. In the case of the internal project, the project manager is aiming to achieve the project aims for improvement of his own employing organisation. In the case of the external project, the project manager's intention is profit from the contract for his employer. The former may well be in a situation where the type of project is unique within the company's history, while the latter is working within an organisation which is structured so as to manage a number of projects in parallel. The project manager in this type of company will be able to rely on much internal expertise and experience in the management of projects. The project manager of an internal project may not have such good fortune, and may also find that functional managers in other parts of the organisation are not totally committed to the project aims.

Where the project is internally generated, it is part of an overall corporate plan, and it competes with other potential projects for the scarce resources of money and expertise. The contractor, on the other hand, is likely to have won the contract in competition with others. Thus a project manager may find himself with a project won by pruning estimates, and in consequence a total budget for the project which makes a profit unlikely from the start. These aspects of projects will influence the project manager's attitudes towards the project, and his behaviour as a manager. They will also influence the options available to the project manager when making decisions about the organisation structure to be employed for the project. Another factor which will influence the shape of the organisation is the people who are available to form the project team.

The co-ordination matrix was employed in the manufacturing industry case, in the multi-project department. Here the project co-ordinator is in a relatively weak position, depending upon his powers of persuasion to obtain commitment from specialists within the functional departments. While the ideas about teams and their leadership still apply, it can be concluded that in this situation the specialist is likely to identify with his functional department, particularly when the leadership there is strong. Where large-scale projects are to be managed in this company, a secondment model is used, setting up a task force under a project leader. This model is also used by the contractor for much of their project work, particularly site operations. The project manager in this situation is particularly powerful.

In the early stages of the project there is usually considerable uncertainty. Is the scheme technically feasible? Can the costs be kept within the budget? Can the project be completed within the timescale? Will the Board of Directors give the go-ahead? These and many other questions will be at the forefront of people's minds. The people involved at this stage are likely to be highly professional in their own sphere. The study of structures and groups showed that at this stage an overlay matrix may be the most obvious choice, with a more democratic style of leadership. However, it was also clear that the project manager must keep control over this phase, ensuring that the designers recognise the implications of their decisions on later phases of the project, and that they do not lose sight of the time, cost and performance objectives.

At later stages in the project, whether erection or construction, a secondment matrix appeared more appropriate on larger projects. However, once seconded, the actual organisation adopted for this phase is likely to be either the functional or the regional type. However, the nature of project work is such that it is essential that the organisation does not become over-bureaucratised. High levels of communication, flexibility, responsiveness and rapid decision-making are the characteristics essential in project management teams. The organisation complexity at this phase, which results from the considerable numbers of organisations and people involved, requires a more autocratic style of management than could be permitted at the design phase. The project manager may well adopt a fairly participative style with his group of senior subordinates, but then require a fairly autocratic style with others involved.

The project manager's role extends beyond that of managing the teams. Another key role is that of ambassador, in which he represents the project to the world around. Whether this is to Board members or clients, this is a vital role which can have considerable bearing on the success of the project. This requires the project manager to be politically sensitive, knowing what to hide and when. The maintenance of confidence in the ability of the team to perform and the protection of that team from outside interferences are both important elements of this role.

Finally, the project manager's role was studied as a model to be emulated by others. If the project manager is particularly successful, others will examine his operating style and mould their own behaviour on his. The project manager thus has a key role to perform in the development of future generations of project managers.

3.6   DISCUSSION QUESTIONS

1. What is meant by a Product Organisation and a Functional Organisation? Discuss how the hierarchical structures in each of these organisations contrast with a matrix management structure.
2. Distinguish between the following types of matrix structure and identify their relevant advantages and disadvantages: co-ordination model, overlay model and secondment model. Provide examples where each type of structure would be appropriate.
3. What are the prerequisites for ensuring the satisfactory use of a matrix management system? How do informal systems complement the formal management systems?
4. Why is it important to define the Project Brief, and what should be covered by this brief?
5. What makes the management of design both particularly difficult but especially important? How can the risks introduced at this stage be minimised?
6. Identify the attributes needed of a project manager under the following headings: technical skills, personal skills, personal attributes and commercial skills.
7. How should a project team be formed and controlled if it is to be as effective as possible? How can this effectiveness be measured, in terms other than the quantity or quality of the work produced?

Part II

Legal Aspects

Chapter 4
Contractors and Contract Law

4.1 INTRODUCTION

Almost inevitably projects will call upon resources beyond those of the organisation that instigates a project. Indeed a company is not likely to remain competitive if it can cater for every eventuality from its own resources. Consequently, most projects will make use of contractors to undertake part or all of a project, according to terms that have been agreed in a contract.

The obvious disadvantage of using contractors is that their first loyalty is to their own firms, and not to the organisation that has sponsored the project. In this context, loyalty will often take the form of trying to maximise the contractor's own profits. This leads to the different forms of contract described in the next section. Once the form of contract has been decided, then possible contractors can tender for the contract. Contractors naturally want to try and meet a client's requirements in terms of their existing practices or products; obviously this may not be in the best interest of the client. Contractors may even decide that it is in their best interest to submit a proposal that is different from the client's specification. Some of the criteria for deciding which tender to accept are presented in section 4.2.2. This preliminary section concludes by identifying the need, and suggesting the means of obtaining the best performance from contractors.

Since the contract is a legally binding document, a knowledge of contract law will help keep problems under control. For example, if the work carried out by a sub-contractor subsequently needs rectification, and the sub-contractor is no longer in business, the main contractor will be liable for the cost of rectification. This and other issues are answered in Chapter 5, sections 5.3–5.7, on Contract Law.

4.2 CONTRACTS AND CONTRACTORS

4.2.1 Forms of Contract

Many forms of contract exist, but they are essentially variations on two types: fixed cost and cost plus. A *cost plus contract* is when the contractor charges for the work done, and the profit is agreed as either a lump sum or a percentage of the total cost. The advantages and disadvantages of each type vary with the nature of the work in the contract.

Fixed price contracts are most appropriate when the project or task is well-defined. This enables contractors to prepare realistic quotations; indeed if the project was ill-defined, no contractor may be willing to tender. During a fixed price contract, the contractor will be intent on maximising his total profit, not just the profit on the contract in question. This can lead to less-profitable contracts being drawn out, although this may make the contractor liable to compensate the client for any resulting loss. A fixed price contract may also inhibit changes to the project, since any changes to the contract will involve an extra payment to the contractor. Indeed, a contractor may be relying on changes in the contract to make any profit at all. In a fixed price contract the contractor may become insolvent, thus leading to a partially completed contract and complicated litigation.

Another disadvantage of the fixed price contract is that it tends to extend the timescale. The task has to be accurately specified before inviting tenders; the contractors need time to prepare their tenders, and throughout the duration of the contract the contractor will be trying to save money not time. While it is possible to have time penalty clauses they are difficult to enforce, not least because some changes to the project are almost inevitable, and the contractor can use this as an excuse for delays. In a fixed price contract the client will also need to supervise the work, to ensure adequate standards and the use of suitable materials.

Thus for fixed price contracts it is in the interests of both the client and the contractor to have a fully and accurately specified task.

The cost plus form of contract reimburses the contractor for all the direct and indirect costs incurred in a contract, and provides an additional payment for profit. The profit element is either a fixed sum or a percentage of the total cost. This type of contract is appropriate when the contract cannot be adequately defined, either because it is development work or because time is important and the contract needs to be let as quickly as possible. Time is saved on tendering, and the contractor should make neither a large profit nor a loss.

The obvious disadvantage of the cost plus form of contract is that it provides little incentive for the contractor either to control the project costs or to meet completion dates. In reality, most contractors seek to maintain a good reputation, and wish to remain eligible for subsequent contracts.

Another disadvantage of the cost plus form of contract is that there are no restraints on the sponsor of the contract to freeze the design or specification. This leads to 'creeping improvements'—that is, minor changes that lead to marginal improvements but at great expense in terms of both the timescale and cost. Of course, creeping improvements should be avoided in any project.

With cost plus contracts a monitoring procedure needs to be agreed between the client and contractor. This is essential if the project is to be kept on course at a reasonable cost. However, such monitoring can be seen as interference, but it is essential if the client is to maintain control over the project's commercial viability.

There are, of course, forms of contract that attempt to attain the advantages of both fixed price contracts and cost plus forms of contract, without the disadvantages. Examples of this are fixed price contracts with:

(1) *Escalation*—for lengthy contracts the cost may be linked to an appropriate index, such as interest rates, materials costs or labour costs.
(2) *Redetermination*—if large contingency sums have been built into the contract.
(3) *Incentives*—the target cost and target profit are agreed, along with a ceiling price and an adjustment formula; this approach is illustrated in the following example.

Example of a Fixed Price Incentive Contract

The following have been agreed as part of contract:

Target cost	£200 000
Target profit	£ 20 000
Ceiling cost	£275 000

The cost and profit sharing arrangement is as follows:

(1) if the target cost is exceeded, the customer pays 75 per cent, and the contractor pays 25 per cent of the excess;
(2) if the cost is less than the target cost, the customer receives 80 per cent of the saving and the contractor receives 20 per cent of the saving.

Case (a) The project cost has an over-run of 25 per cent, that is, £50 000.

The customer pays:	£	£
Target cost	200 000	
75 per cent of £50 000 over-spend	37 500	
The target profit	20 000	
Total project cost		257 500

The contractor receives:
The target profit   20 000
less 25 per cent of the
     £50 000 over-spend   12 500
Contractor's profit   7 500

The contractor's profit margin is now 2.9 per cent (7 500/257 500), as opposed to the intended profit margin of 9.1 per cent 20 000/220 000).

Case (*b*) The project has cost £20 000 less than expected.

The Customer pays:	£	£
Target cost	200 000	
Less 80 per cent of £20 000, saving	−16 000	
The target profit	20 000	
Total project cost		204 000
The Contractor receives:		
The target profit	20 000	
and 20 per cent of the £20 000 saving	4 000	
		24 000

The contractor's profit margin is now 11.8 per cent (24 000/204 000). These figures can be compared with the target total cost of £220 000 and a profit margin of 9 per cent.

The advantages and disadvantages of the different forms of contract are summarised in figure 4.1. Obviously, the more closely specified the contract, the more likely it is to be offered on a fixed price basis, with the acceptance based on minimum cost. The use of the different contract types is discussed further in Chapter 7, section 7.2, in the context of computer projects.

Even in the most clear-cut case, there may well be a need for variations during the course of a contract. In order to maintain close control on any changes, it is often desirable to have only one person with authority to vary the contract terms (the project manager, or sometimes the *contract manager*). The form of words in the contract may be as follows '. . . the contractor shall carry out such additional work up to 15 per cent of the contract price, as is instructed by the client at a mutually agreed price'.

The terms of the contract are, of course, very important and many organisations have their own standard contracts; in addition, professional organisations produce model forms of contract (for example, Institution of Mechanical Engineers, Institution of Electrical Engineers) to cover different

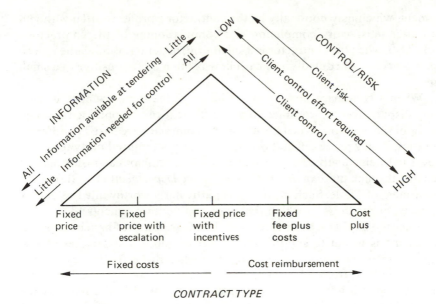

CONTRACT TYPE

Figure 4.1   Characteristics of the different contract types

types of work. An important aspect is for the provision of arbitration, should there be a dispute concerning the contract.

Just as clients will use standard forms of contract, the contractors too may provide a quotation that is subject to their own terms and conditions. These terms and conditions may contradict those of the client, and this can lead to a 'battle of forms'. Obviously the terms associated with a contract need to be studied carefully, and it may be necessary to obtain legal advice. In general, when proposals and counter proposals are being made, it will be the last counter proposal that governs the terms of the contract; this aspect is discussed more fully in section 4.3.2. Wearne (1984) provides a survey of some different model forms of contract, and also a résumé of some of the problems that occur in contracts and how these problems might be resolved or eliminated.

The terms of payment will be specified in the contract. In large contracts, stage payments will be made according to a prescribed scheme. Contractors will want a payment with the order that will cover the '*front end*' *costs*—these are the costs that the contractor will have incurred in gaining the contract. The contractor will also want *progress payments*, and this is likely to be in the interests of both parties. However, the customer or client is going to want completion of specified stages or '*milestones*' to be reached, while the contractor will want payment based on the start of activities. In some cases a more satisfactory arrangement is for an independent valuation of the work to be carried out, and the contractor to be paid accordingly. The final payment, which may notionally be the contractor's profit, is often withheld until after satisfactory completion and commissioning of the contractor's work. Also, when the contractor is a small firm, better terms might be agreed if payments are scheduled, since cash flow is often more sensitive in a small firm.

When a contractor and client are based in the same country, if there is any dispute over the contract then a settlement can be obtained through the law of that country. When the client and contractor are based in different countries, it may prove difficult or impossible for an injured party to obtain redress in foreign courts. For this reason a UK contractor can insure against such loss through the *Export Credits Guarantee Department* (ECGD) of the Department of Trade. Such insurance is particularly worthwhile for lengthy contracts, or for those in which there are no interim payments.

Since the ECGD scheme is run with UK Government financial support, the scheme is meant to support the export of goods and services of UK origin. Very often major UK export contracts will be funded by a fixed interest rate loan from a UK bank to the overseas buyer. These long-term loans can be at a fixed interest rate that is guaranteed to the bank by the ECGD. The contractor will need to check that the terms of the loan are compatible with the contract, notably:

(1) the total amount of the loan must cover the contract cost and allowance for variations, and any cost escalations;
(2) the loan must be able to make funds available to the contractor at the appropriate rate.

These and other aspects of ECGD support are discussed by Brown (1984). The ECGD also runs a *tender to contract scheme* (TTC) that enables firms to quote for a contract in foreign currencies without being made vulnerable to exchange rate changes. Brown (1984) also discusses other sources of international finance and aid for exports.

The contract is likely to state the standards that are to be adopted (for example, British Standards), and in addition some form of quality control will be needed. A compliance to a recognised standard, such as BS 5750 Quality Systems, can provide the customer with the necessary quality assurance. The use of sub-contractors may also be covered in the contract. The client may require the use of any sub-contractors to be approved. Equally, it will be in the interests of the contractor to vet any possible sub-contractor prior to tendering for the main contract.

4.2.2 Criteria for Selecting Contractors

The tender with the lowest cost is not necessarily the one to accept, Warby (1984) quotes John Ruskin as stating that "it is unwise to pay too much, but it is worse to pay too little. When you pay too much, you lose a little money—that's all. When you pay too little, you sometimes lose everything, because the thing you bought was incapable of doing the thing it was bought to do. The common law of business balance prohibits paying a little and getting a lot—it cannot be done. If you deal with the lowest bidder it is as well to add something for the risk you run. And if you do that you will have enough to pay for something better." Some additional criteria that might be important in selecting a contractor include:

(1) Reputation or in-house bias—it may be vital for a contract that is part of a large project to be completed on time, in which case past experience can be a useful guide to the future.
(2) Use of innovative solutions—these may enable time or cost to be cut, without cutting corners.
(3) Health and vitality of the contractor—if the contractor goes out of business, an incompleted contract causes added costs and delay.
(4) Ethics—a contractor may propose methods that would contravene the Health and Safety at Work Act; the client also has a reputation to be preserved.
(5) Conditions of payment—cash flow costs can and should be assessed.
(6) Sub-contractors—if a contractor makes any use of sub-contractors, the contract will be more difficult to control, and is thus more likely to over-run.
(7) Flexibility, timing, the extent of the contractor's resources and location.

Conversely, some engineers will find themselves involved with preparing tenders, and the following considerations may be useful:

(1) find out as much as possible about the company and the task;
(2) assess the competition for the contract;
(3) know your own company's strengths and weaknesses;
(4) produce a correct costing, for both overheads and direct costs;
(5) assess the scope for future associated work that might justify reducing the current quotation;
(6) demonstrate experience and show flexibility;
(7) look for innovative solutions;
(8) offer anything that might be a useful addition;
(9) provide a personal touch;
(10) avoid small margins;
(11) decide if some form of legal protection is needed for any design work that is submitted with a tender.

In addition, the tender document issued by a potential customer needs to be considered very carefully. Warby (1984) discusses these aspects in some detail, and presents the checklist that is shown in figure 4.2.

1.0 SCOPE
Define the responsibilities for the functions below:

Research and development
Design calculations
Specification of equipment, materials and workmanship
Drawings
    arrangement and detail
    parts lists
    supply of prints
Design approval
Material supply
Equipment supply
Receipt, off-loading and storage of material and
    equipment
Manufacture
Jigs and fixtures, special equipment
Finishing
Packing and temporary protection
Loading at works
Transport
Port charges
Sea/air freight
Off-loading and storage at destination
Installation
    supervision labour
    equipment and site facilities
Commissioning
Maintenance
Training
Operating and maintenance manuals
Inspection and acceptance

2.0 EXTENT OF SUPPLY
Define the numbers and types of the following to be
    supplied:

Goods
Spares
Jigs and fixtures which become the purchaser's property
Special equipment which becomes the purchaser's
    property

3.0 SPECIFICATION

3.1 Quality
Quality assurance
Quality control

3.1.1 Performance
Physical dimensions
Equipment list
Static performance
Dynamic performance
Instrumentation
Service conditions
Environment
Working life

3.1.2 Materials
Raw materials
Equipment
Testing and certification

3.1.3 Workmanship
Workmanship
Testing and qualifications

3.2 Delivery
Define the dates applicable for the following:

    Approval of design
    Approval of prototype
    Commencement of deliveries
    Intermediate deliveries
    Completion of deliveries
    Holidays

3.3 Price
Define the following:
    Fixed or escalating
    Contract price adjustment formula if escalating
    Lump sum or unit price for a quantity
    Currency of payments
    Period of validity

4.0 CONDITIONS OF CONTRACT
Define the legal framework for the following:

    Terms of payment
    Variations
    Time for completion
    Damages
    Defects liability
    Accidents and damage
    Determination

Figure 4.2   Tender checklist [Reprinted by permission of the Council of the Institution of Mechanical Engineers from D. J. Warby (1984), 'Preparing the Offer', Proc. I. Mech. E., Vol. 198B, No. 10]

The cost to a contractor of preparing a tender can easily be overlooked. As the value of a contract increases, the cost of bidding falls as a percentage of the project cost. Thus, if a contractor only bids for low value contracts, he has to be very successful in winning contracts if he is to remain in business. Figure 4.3 shows the relation between the cost of bid preparation and the value of a contract for the aerospace industries. Contractors would be well advised to prepare a similar graph for their own business sector. By implication, contractors should not accept every invitation to tender. A contractor should prepare a bid decision score sheet that lists the criteria that influence the decision of whether or not to bid. Each criteria can then be given a rating for the contractor and his main rival, and multiplied by

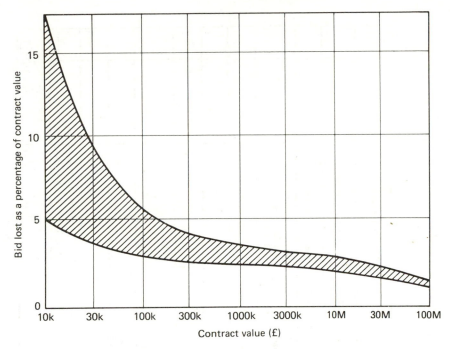

Figure 4.3   Range of costs for bid preparation in the aerospace industries

a weighing factor for the criteria to give an overall rating in the manner shown in figure 4.4. A contractor can then make a decision on whether or not to bid on a rational basis.

If the contractor is proud of his record, he is likely to submit a list of recently completed contracts with some supporting documentation. Equally if the client is considering a contractor for the first time, a list of recently completed contracts should be asked for, and checked out.

Care must also be taken if sub-contractors are to be used. Some contracts will not permit the use of sub-contractors unless sanctioned by the client. Other projects may be so large that they require a hierarchy of managing contractor, main contractors and sub-contractors. With such a system the responsibilities have to be very accurately specified.

To prevent a supplier or contractor causing a delay, he may often need to be urged both frequently and strenuously. Often it can be advantageous for such progressing to be carried out by the purchasing or buying department of a customer; in this way the supplier or contractor will realise that future business may also be at stake.

4.2.3   Contractor Motivation

Contractors are in business to make a profit, and their first loyalty is to

Bid criteria	*Bid criteria weighting factor*	*Bid criteria score* — *Positive* (10, 9, 8, 7)	*Bid criteria score* — *Neutral* (6, 5, 4, 3)	*Bid criteria score* — *Negative* (2, 1, 0)	*Bid criteria rating (weighting factor × score)* — *AJ Contractors*	*Bid criteria rating (weighting factor × score)* — *Main Rival*
1. Previous experience of similar work		Strong in-house experience	Average in-house or imported	Weak or new area		
2. Client rapport		Long-standing and successful collaboration	Occasional previous work	No previous contact		
3. Human resources		Good and available	Good but not available	Available but not good		
4. Facilities (physical resources)		Available and/or favourably located	No impact	No spare capacity nor new facilities required		
5.						
6.						
7.						
8. Competition		AJ is sole source	Open or unknown	Rigged for rival		
9.						
10.						
OTHER COMMENTS:				TOTAL SCORE (Σ Bid criteria ratings) MAXIMUM POINTS AVAILABLE	______	______

DECISION:

BID ☐ NO BID ☐

Figure 4.4 A bid decision score sheet for AJ Contractors

themselves, not the client organisation. Jay (1967) identifies this problem of loyalty, and compares contractors to mercenaries. Mercenaries (or contractors) are liable to desert to another prince (contract) or even to an enemy (competitor) if the pay (profit) is better. Contractors are also subject to their own industrial troubles, and if they manufacture a successful product under contract, they may wish to market it themselves.

A powerful means of motivating contractors is to commit them to a project's success. To this end it is necessary to familiarise them with the whole project, and how their contribution fits into the overall scheme. By making contractors feel involved and important, they should then feel responsibility and have obligations towards a project's success.

4.3 GENERAL PRINCIPLES OF CONTRACT

The two vital questions to be asked when considering the legal aspects of a contract are these: Have the parties made a contract? If so, on whose terms of business? Unless the supplier and the customer have entered into a binding contract, each party is still free to change his mind. So the supplier could withdraw a quotation or change his terms, for example, by raising his prices, at any moment until the quotation has been unconditionally accepted by the customer. Similarly the customer is under no obligation to accept a quotation, even though long and expensive negotiations may have preceded it, and considerable costs have been borne by the supplier in preparing a tender with detailed specifications and drawings.

The customer may have gone as far as to send a letter of intent to the supplier, but this does not normally create a contract between them. It is simply an indication that the customer is likely to place an order with the supplier rather than a legal commitment. The result is that if the supplier does not deliver equipment on time, the customer will have no redress and will be unable to recover compensation for breach of contract, since no contract exists between the parties.

Assuming the parties have made a contract, if a dispute later arises between them, it will be crucially important to discover whether the contract was made on the supplier's conditions of sale or the customer's conditions of purchase. Their standard terms are likely to be very different, as each party views the arrangement from a different perspective. Among the variable factors are these: Is the price quoted a fixed price, or has the supplier included a price variation clause? Does the price include the cost of carriage to the customer's site, or is it ex works? Is the risk of accidental damage during transit to be carried by the supplier or the customer? When is the ownership in the goods to pass to the customer? Will the supplier be responsible for defects in the goods or their failure to match the specification precisely, or has such liability been excluded by a disclaimer clause? Will the customer have a right to compensation if the goods are delivered late or the installation is not completed on time, or has he signed this right away by making the contract on the supplier's terms? The following sections provide answers to these questions.

4.3.1 Offer and Acceptance

Let us consider first whether a contract has come into existence. A *contract* may be defined as a bargain between the parties, whereby one party agrees or promises to do something (such as to supply equipment) in return for the promise of the other party (usually, to pay the price). To decide whether an agreement or contract exists—and the words 'contract' and 'agreement' are used in this chapter as synonyms—look at all the circumstances to see whether one party has made an offer which the other party has accepted. The test is an objective one and in reaching a decision one must consider what the parties said, wrote and did. It is important to note that very few contracts in English law need to be made in writing, and that commercial contracts may be oral, written or by conduct. So an agreement made on the telephone is as binding as one made by an exchange of documents. However it is, of course, much more difficult to prove in court a contract made by word of mouth, even though this is simply a matter of evidence— that is, can the party making the allegation prove what was said. What is irrelevant is the subjective view of either party, as no doubt each believes the contract to be made on his own terms or to be non-existent.

What is an *offer*? It is a statement that the person making it is willing to contract on the terms stated. It should be distinguished from 'an *invitation to treat*', such as the display of goods in a shop window or in catalogues, where the supplier is inviting the customer to make an offer to him which he may reject or accept as he thinks fit. An *acceptance* will result in a contract, if it is in precisely the same terms as the offer without any qualification or addition. In short, the person receiving the offer (the offeree) responds with a simple 'Yes', not 'Yes, provided. . .'. As soon as the offeree purports to introduce a change, this is treated as a *counter offer*, not an acceptance; the effect is that the counter offer rejects and kills the original offer once and for all, and is itself capable of acceptance by the original offeror. We shall see shortly how this game is played between suppliers and customers in the so-called 'battle of the forms'.

Normally an offer may be '*revoked*' or withdrawn at any moment prior to acceptance, unless an *option agreement* provides for it to be kept open for a fixed period. This is why the supplier or customer is free to change his mind until the contract is made by acceptance of the offer. When is an acceptance effective? The general rule is that a contract will not be created until the acceptance is communicated to the offeror. Where acceptance occurs by an almost instantaneous method of communication (for example telephone or telex), no contract will be made until the acceptance is received. However, an exceptional rule applies where the parties are dealing through the post; here the contract is made when and where the letter of acceptance is posted, so that even if it is lost in the post, still the parties will be bound.

4.3.2 The Battle of the Forms

Not surprisingly, in the case of major contracts for the supply of goods, or for the supply and installation of capital equipment, the parties do not normally commit themselves only by word of mouth. Usually documents pass from one to the other, with each party's conditions hidden away in small print on the back. The sequence of documentation is often as follows:

(1) the customer sends an enquiry form to the supplier;
(2) the supplier replies with a quotation;
(3) the customer places an order and at the same time he may enclose an acknowledgement slip for signature and return by the supplier;
(4) the supplier returns the acknowledgement slip.

In the above typical sequence the legal events are as follows:

(1) an invitation to treat by the customer;
(2) an offer by the supplier;
(3) a counter offer by the customer;
(4) an acceptance by the supplier—the contract is made at this stage on the customer's terms.

However, not every 'battle of the forms' is as straightforward as this. There are many variations in tactics. For example at stage (4) above, the supplier, instead of returning the customer's acknowledgement slip, may (and should) send his own acknowledgement of order setting out his conditions of sale. In short, the party wins the battle who 'fires the last shot'—that is, whose piece of paper is the last to be sent off.

4.3.3 Form and Contents—Terms

We have briefly considered the importance of ascertaining on whose terms the contract has been made. These may be oral, written or a combination of both. It is not enough simply to look at the documents, for there are likely to have been lengthy negotiations including perhaps optimistic statements and promises by the supplier. Do such statements form part and become terms of the contract? *Terms* of the contract (that is, provisions forming part of the contract itself) must be distinguished from mere *representations* (that is, preliminary statements made during negotiations which induce the making of the contract but are not part of it). Into which category a statement falls depends on intention, again viewed objectively; if made by a person with special expertise compared with the other, at or shortly before the contract was created, it is more likely to be a term.

Clearly it is dangerous for a supplier during negotiations to make glib statements about delivery or performance of equipment, especially when those statements are repeated just before the written documentation is exchanged, as this is likely to result in such statements becoming part of the contract. This will be useful for the customer if he later seeks compensation for late delivery, or failure to meet the performance criteria; for the supplier will have broken the contract and will be liable in damages to the customer. If the only *remedy* sought by the customer is *compensation*, it is enough to prove that a term of the contract has been broken and caused him loss. If, however, the customer wishes to *terminate* the contract because of the supplier's default, it is necessary to look more closely at the type of term broken and the severity of the breach.

Historically, contractual terms have been classified into two categories: conditions and warranties. A *condition* is a term of major importance; its breach entitles the innocent party to cancel the contract—that is, to treat the contract as ended and to refuse to perform his own obligations, and in addition where appropriate to claim damages. A *warranty* is a minor, subsidiary term; its breach gives only a remedy of damages. Often it is difficult to work out which type of term has been broken. Sometimes the wording of the contract itself will resolve the difficulty. It is possible to draft a contract so as to make clear from the start into which category terms are to fall, but the use of the words 'condition' and 'warranty' is not conclusive, as the words are not always meant to have a technical meaning, and in particular are generally not used in a legalistic way by businessmen. In such cases where the importance of the term is unclear, it is described as an *'intermediate stipulation.'* This is the modern judicial approach. In such cases the innocent party's remedy will depend on how seriously the contract is broken—a grave breach is equivalent to a breach of condition, a minor breach to a breach of warranty. So termination is available only for a serious breach.

4.3.4 Misrepresentation

In contrast to a term, a *misrepresentation* is a statement made by one party to the other, before or at the time of contracting, with regard to some existing fact, which is one of the causes inducing the contract and proves to be untrue. Statements of intention or opinion are not misrepresentations. Nor is silence—there is generally no duty of disclosure, although an important exception relates to the law of insurance where the proposer must not merely answer all questions truthfully, but must in addition reveal unasked to the insurer all other material facts.

There are two main types of misrepresentation: fraudulent and innocent. *Fraudulent misrepresentation* is a false statement, made knowingly or
without belief in its truth. Anything else is *innocent misrepresentation*, even
if it is made carelessly.

The remedies for misrepresentation and for breach of contract, including damages, are discussed more fully in section 4.6.

4.3.5 Privity of Contract

We have referred frequently to 'the parties to the contract', but there may
be other businesses involved in its performance. For example, a contractor
may purchase equipment from a manufacturer which the contractor will
install as part of a project for the client. The manufacturer may offer a
'*guarantee*' with the equipment for the benefit of the end user and frequently
the guarantee will limit the manufacturer's responsibility to defects occurring
during the first 12 months. What happens if the equipment installed by the
contractor proves faulty outside the guarantee period? Will the manufacturer's guarantee protect the contractor? A third party may also become
involved by way of sub-contracting. If the sub-contractor does not meet the
specification in the main contract or his workmanship is shoddy, will the
client's redress be against the main contractor or the sub-contractor?

These problems can be resolved by reference to the *privity of contract
rule*—only parties to a contract can sue or be sued upon it. Thus a third
party cannot claim the benefit of a provision in another's contract and for
this reason a contractor cannot shelter behind a limitation clause in a
manufacturer's guarantee. Similarly a sub-contractor is not liable in contract
to the client, and the main contractor continues to be responsible for the
quality of the equipment and workmanship. Of course, if the main contractor
is held liable in damages to the client for breach of contract, he will himself
have a right of action (an '*indemnity*') against the sub-contractor on the
same basis.

4.3.6 Frustration and *Force Majeure*

So far we have concentrated our attention on the initial stages of the contract
and the circumstances surrounding its formation. What happens if those
circumstances change unexpectedly after the contract has been made, but
before it has been completed? This may happen because of a Government
embargo on the import or export of equipment or requisitioning by the
Government during wartime. Here the problem does not involve breach of
contract. Each party is keen to perform his obligations but through no fault
of his own is unable to do so because the situation has changed. It would
seem unreasonable to penalise a party for non-performance in such circumstances. The law adopts a sensible view here, and by the doctrine of

'frustration', treats the contract as discharged and at an end, from the moment of the frustrating event—it is a valid contract until that moment. The broad effect is that if advance payments have been made to the contractor, he may retain these to the extent that he has incurred expenses in carrying out the contract so far; subject to that the employer's liability to pay disappears.

Nevertheless, the operation of frustration has two major disadvantages. Firstly, it is often difficult to decide at the time when the unexpected event occurs whether the change is so fundamental as to make further performance either impossible or radically different from the commercial undertaking on which the parties originally embarked. Secondly, the effect of frustration, as we saw, is to terminate the contract: it does not permit a contractor to delay completion with impunity. Either the contract is off or it is on. For this reason, commercial contracts frequently include so-called *force majeure* clauses, which relieve the contractor from liability and allow an extension of time if completion of the job is delayed 'through circumstances beyond his reasonable control'. Sometimes such clauses add to this general formula a list of particular dangers, such as strikes, which may arise.

4.3.7 Agency

The last topic to which we shall refer in this brief survey of contractual principles is *agency*. Although its significance and ramifications are so wide that complete textbooks are devoted solely to the law of agency, we shall highlight merely a few aspects.

Clearly when a company enters into a contract, as it is an inanimate legal person, it must do so through the agency of its employees. Suppose that S Ltd, a supplier, is negotiating a contract with C Ltd, a customer. Considering this from the point of view of C Ltd, let us assume that the company has an internal arrangement whereby all contracts should be finally and formally tied up by its purchasing department. X is employed in that department as a senior buyer and is authorised to make contracts for C Ltd up to a figure of £100 000. Here C Ltd is the *principal*, X is the *agent*, and S Ltd is the third party.

No difficulty arises where the agent has *actual* authority to act as he does. Actual authority may be express or implied. In the above example if X orders goods costing £75 000, the contract is binding on C Ltd because X has *express* authority to enter into contracts up to £100 000. (If X's authority has not been specifically clarified in his contract with C Ltd, then he will have *implied authority* from his position as a buyer to make contracts within the usual authority of a buyer.)

However, the position will be less straightforward where X as agent for C Ltd makes the contract with S Ltd for, say, £120 000. We know that as regards the internal arrangements with his employer, X has exceeded his authority and, if it were not for some additional rule, the contract would not be binding on C Ltd at all. But why should S Ltd be prejudiced when without knowledge of the internal limit on X's authority they have made a contract with C Ltd? This is covered by what is known as 'apparent' or 'ostensible' authority: if the agent appears to third parties to have authority, and this appearance results from the words or conduct of the principal by appointing him to a particular post, the principal is bound by the contract even though the agent exceeds his actual authority. (This is sometimes described as agency by 'estoppel'—that is, the principal is estopped or prohibited from denying that the agent is properly authorised.) The final point to note is that only the third party benefits from this rule; as between the principal and the agent the act is unauthorised, the agent has broken his contract with the principal and the agent could be called upon to compensate the principal for any resulting loss.

Another situation where an employee without actual authority may bind his employer vis-à-vis a third party occurs where an engineer, who has been negotiating a contract with a prospective supplier or customer, goes so far as to finalise the contract, even though his company may have a structure whereby all such contracts are to be made only by the purchasing department. Again, although the engineer may be acting against instructions in so doing, unless the third party is aware of that fact, the employer will be committed. In theory this problem can be overcome by the employer notifying all prospective suppliers or customers of the limitation on the authority of particular employees, but in practice this is an unrealistic solution.

4.4 SUPPLIERS' DUTIES—GOODS

We have seen that a supplier may take on particular contractual responsibilities by express statements, written and oral, which may be either terms of the contract or misrepresentations. Such express statements, however, are not the sum total of the supplier's obligations; for additional terms may be implied into the contract in a number of ways, such as by custom, and in particular by legislation. We now turn to these implied statutory duties. They are limited in scope, and can be viewed as imposing on a supplier a narrow range of obligations which would fall within the reasonable expectation of a customer in a commercial contract. For example, when businessmen are discussing the proposed sale of component parts, it is unlikely that they will bother to discuss whether or not the seller is the true owner of the components, and what will happen if they turn out to be stolen—that is, it is unlikely that there will be an express term dealing with this facet of the agreement. Fortunately this gap is filled by terms implied by the Acts to be discussed.

When dealing with the implied terms, in most cases we have no difficulty in deciding whether they are 'conditions' or 'warranties'—as we have seen (section 4.3.3) an important division when considering the remedy of the customer—as the Acts label the terms for us. Most are described as conditions, a few as warranties, and only exceptionally are they described as 'terms' when we can discover the customer's remedy only by considering the severity of the breach.

4.4.1 Types of Supply Contracts

Contracts for the *sale of goods* are governed by the Sale of Goods Act 1979 ('the 1979 Act'). This will apply to the supply of component parts, end products and any goods where the seller's obligation is not more extensive than making the item and delivering it to the buyer.

Where the supplier agrees to supply equipment and install it, the contract is classified as a contract for '*work and materials*', not sale of goods. The essential distinction is that a contract for work and materials involves not merely the transfer of goods but also the *supply of services*, as skill and labour by the supplier is an integral part of the contract. Examples of such contracts include the supply and installation of capital equipment, building and construction contracts, and contracts for maintenance and repair. Such contracts have two elements: (1) the supply of goods and (2) the supply of services. This is clearly shown in the well-known judicial definition: 'half the rendering of service and, in a sense, half the supply of goods'. Not surprisingly, these contracts impose two different sets of obligations on the supplier, dealing separately with the goods' element and the services' element. The relevant legislation here, whose title reveals this double set of obligations, is the Supply of Goods and Services Act 1982 ('the 1982 Act').

Another method frequently used in business which enables the customer to have the use of goods, though not their ownership, is the contract of *hire*. There are various commercial types of hire—rental agreements, contract hire, equipment leasing, finance leasing. In law they are all treated in the same way. The common factor is that the customer (usually known as the '*bailee*' or '*hirer*') is not entitled to become the owner of the goods under the terms of the contract but merely has the right to the use and possession of the goods during the contract period. The supplier (usually called the '*bailor*' or '*owner*') retains the title, property and ownership throughout and ultimately becomes entitled once more to possession too. Contracts of hire are also covered by the 1982 Act.

4.4.2 Strict Liability

One crucial concept needs to be grasped immediately. Where a supplier is *strictly liable*, for example, for defects in the goods, this means that he is responsible whether or not he knew, or could have known, of the breach of contract. For example, if the goods are defective, it is no defence for the supplier to be able to prove that the defect was a latent defect, and that its presence could only have come to light after the goods had been used for a substantial period. In other words, the customer does not need to prove that the supplier was negligent in supplying goods in breach of contract.

Many of the cases on this point which have come before the courts relate to consumer products which have been faulty and caused personal injury to the buyer. For example, in one notable case a doctor bought a pair of underpants which caused severe dermatitis; they contained an excess of sulphites which was invisible to the naked eye; even so the shopkeeper was strictly liable to the consumer for his personal injury. In other cases, retailers have been held responsible for a stone in a bun which broke the customer's tooth, a hair crack in a catapult which shattered and blinded a child in one eye, and tuberculosis germs in milk.

The law on this point is the same whether the customer is a business customer or a private consumer. So if a contractor were to buy in components from a reputable supplier, check a sample for quality, install the components into his end product and then supply and install that equipment for a customer, the contractor would be liable to the customer if equipment failed because of a faulty component. This is so, even though the contractor was not negligent in that his supplier was reputable, and his quality control was careful. Many retailers and contractors think that this is an unreasonably harsh rule, but it should be remembered that they can pass the buck back to their wholesalers, manufacturers or component suppliers by virtue of the contract made between them, which includes similar strict obligations.

4.4.3 Description and Specifications

Where goods are supplied by description, both the 1979 and 1982 Acts imply a condition that the goods ultimately delivered should correspond with the description. The provision applies to both oral and written descriptions. It is particularly important where the parties agree in advance on the materials and equipment to be used, and set out their requirements in a detailed specification. The supplier must comply with all the elements in the description, including the size, quantity and method of packing. The House of Lords has stressed the importance of the supplier adhering precisely to the contract description. It is clearly no defence for the supplier to prove that, although he has delivered something different, it is of the same quality and fulfils the same needs as the article agreed upon. As one judge expressed it, 'If the article they have purchased is not in fact the article that has been delivered, they are entitled to reject it, even though it is the commercial equivalent of that which they have bought'. So if the supplier wishes to reserve to himself the right to deliver an equivalent item instead, his conditions of sale must contain such an express term to protect him. Such a modifying clause is common in practice.

4.4.4 Quality and Fitness

The basic rule in supply of goods cases is *caveat emptor*—that is, 'let the buyer beware'. In other words, it is for the customer to make up his own mind as to whether the goods ordered are of the right quality and suitable for his requirements. However, this basic rule is substantially eroded where the supplier acts 'in the course of a business'. So where the customer is contracting with a commercial supplier, he has the benefit of two implied terms relating to quality and fitness.

Merchantable Quality

The first of the two implied terms—both are conditions—is important where the equipment proves to be defective in ordinary day-to-day use. The obligation is that the goods must be of '*merchantable quality*'. The statutory definition of this expression is that the goods 'are as fit for the purpose or purposes for which goods of that kind are commonly supplied as it is reasonable to expect having regard to any description applied to them, the price (if relevant) and all other relevant circumstances'. Broadly this means that they are reasonably fit for the purpose for which such goods are normally used. The emphasis is on whether or not the goods function properly. So if equipment is unsafe or fails to operate because of a defect or fault, the supplier will be liable.

A frequent problem arises where equipment seems satisfactory when first delivered and commissioned, and causes no problem for perhaps many months. Is the supplier liable for a later failure? Although the Acts do not make it clear beyond doubt that the condition as to merchantability also covers *durability*, decided cases make it clear that this is implicit—that is, the goods must remain usable for a reasonable period after delivery. What is a reasonable period is a difficult question of fact. In the case of a dispute it will usually be necessary to consult an expert, to see whether equipment of the type in question, bearing in mind the price and the related quality, should have developed a fault in the particular component at that stage in its life. Generally the higher the price of a product or component, the higher the standard of merchantability to be expected by the customer.

Often complaints of this type involve a reference to the manufacturer's guarantee. A number of points should be noted.

(1) If the manufacturer has not supplied the customer direct, then the guarantee cannot protect the supplier (see privity of contract, section 4.3.5).

(2) Even if the manufacturer is in a direct contractual relationship with the customer, the guarantee (usually a clause in the manufacturer's conditions of sale) will not form part of the particular contract where the customer has won the battle of the forms (section 4.3.2).

(3) Where the equipment has failed after the expiry of the guarantee period, the customer can argue that the guarantee is an unreasonable exemption clause and struck down by the Unfair Contract Terms Act 1977 (see section 4.7).

The 1979 and 1982 Acts contain *two exceptions* where the condition is not implied.

(1) The first exception relates to 'defects specifically drawn to the buyer's attention before the contract is made'. This is unlikely to apply to a normal project, since it is relevant only where the equipment is complete before the contract is made. For example, a defective clutch in a secondhand car pointed out to the buyer before the purchase is agreed will not be the responsibility of the seller.

(2) The second exception applies where the customer 'examines the goods before the contract is made, as regards defects which that examination ought to reveal'. Again this is unlikely to matter in the context of project engineering, for once more, equipment needs to be completed before the contract is made, and the buyer examines it at that stage. Where the exception does apply, the customer cannot complain about obvious defects which his examination should have picked up.

Fitness for Purpose

The condition of fitness for purpose imposes a more stringent responsibility on the supplier but applies in narrower circumstances than the merchantable quality condition. Where it applies, the goods supplied must be fit for the '*particular* purpose'. Clearly it would be unreasonable to saddle a supplier with this liability, unless the supplier is made aware of the special requirements of his customer before closing the deal. That point is catered for in the Acts by providing that the fitness term catches the supplier only where the customer 'makes known . . . any particular purpose for which the goods are being acquired'.

This provision is significant where the customer's purpose is unusual, special and exceptional; for then it will be useless for him to obtain equipment which satisfies the functions for which others normally require it, if the equipment does not match his special needs. Suppose that a main contractor has agreed to erect a plant in an area with unusual climatic conditions—perhaps extreme temperatures, as in the Gulf or the Arctic, or the dusty conditions of Turkey, the humidity of a tropical area or the extreme stresses of an off-shore oil production platform. The main contractor intends to sub-contract the fabrication and manufacture of part of the plant. It will be in his interest to notify the sub-contractors of the special circumstances in which their equipment is to be used so as to incorporate into the sub-contracts the implied term of fitness. If then the plant fails to work properly, because of the failure of some of the sub-contractor's equipment, the sub-contractor will be liable to the main contractor (himself liable to the employer). The mere fact that the equipment would have performed satisfactorily in normal site conditions, in the United Kingdom or some other temperate climate—and so proves to be of merchantable quality—will not provide the sub-contractor with an excuse.

Does this mean that a contractor should not tender for a job in an area which is novel for him? His solution is to make it clear to the prospective customer that he has no experience of working in such conditions and that, though his equipment is reliable in normal circumstances, he cannot be sure that it will perform properly on the contract site. This way out is permitted by the statutes in that they provide that there is no fitness condition implied 'where the circumstances show that the buyer does not rely, or that it is unreasonable for him to rely, on the skill or judgment of the seller'.

Similar problems occur in the building and construction industry. The building contractor will supply building materials and erect a structure in compliance with instructions and specifications given to him by his client or his client's professional adviser. Nevertheless the contractor will still be responsible if the materials are not of merchantable quality, and are unfit for their normal purpose. He will not be liable, however, if the materials are unsuitable because of special site conditions, since presumably the client will have relied on himself or his advisers in choosing the design and materials for the particular circumstances.

4.5   SUPPLIERS' DUTIES—SERVICES

The *provision of services* may be placed in two broad categories. The first group comprises pure services, ranging from cleaners, carriers and warehousemen to professional services from architects, structural engineers, quantity surveyors, accountants and solicitors. The second group comprises contracts where the provision of the service is tied to the transfer of equipment or materials—that is, contracts for work and materials. We now turn to contracts for services, and the services or work element of a contract for work and materials. These are partly covered by the Supply of Goods and Services Act 1982 which deals with just three aspects—the duty of care, time and charges.

4.5.1   Duty of Care

The obligation of the supplier of a service to 'carry out the service with

· reasonable care and skill' self-evidently does not impose strict liability on the supplier. For example, if an engineer designs a structure and he does not fall below the standards of professional competence exercised by a member of his profession, he will not be liable if the structure collapses, for he has not been negligent. This contrasts with the strict liability of a supplier of goods considered in section 4.4.2.

That is the limit of the implied obligation. But a supplier may expressly take on a higher obligation, and there have been a number of cases where suppliers have in effect promised to reach their objective, and to achieve the desired result. Thus in the above example, if the structural engineer promised not merely to do his competent best in designing the structure but to design a structure which would be safe and would not collapse, the latter express undertaking would bind him and make him responsible for the failure of the structure.

4.5.2 Time

The Act is also concerned with the *time for performance*. However, it is limited to the situation where the parties have not agreed when the service is to be carried out. The Act provides that, where the time for the service to be carried out is not fixed, there is an implied term that the supplier will carry it out within a reasonable time. This is always a difficult question of fact.

Where the time is fixed, the relationship between the parties is governed by the common law. Normally, in a commercial contract, time is considered to be '*of the essence*'—that is, it is an important term that the supplier will deliver or perform by the specified date, otherwise the customer will be entitled to cancel the contract and to recover damages. (If, of course, the delay is caused by circumstances beyond the control of the supplier, *force majeure* clauses need to be borne in mind.)

4.5.3 Charges

Again in the case of costs and *charges*, the Act only interferes in the absence of an express agreement between the parties. So if a firm quotation for the work to be done has been made and accepted by the customer, that price will be the contract price and there will be no room for the Act to operate. However, where the price has not been fixed between the parties, the Act implies a term that the customer will pay a reasonable charge. This will not help, then, where a customer has agreed an excessive charge and so has made a bad bargain. But it will be particularly helpful where a trader is called in to carry out maintenance or to do emergency repairs, without the customer remembering to fix the hourly rate or total sum payable in advance. The same applies to professional services.

Finally, it should be mentioned that the use of the words 'quotation' · or 'estimate' does not in itself show for sure whether it is a contract for a fixed sum or not. As a rule of thumb, the word '*quotation*' has an increasing tendency to mean a firm offer by the supplier to do the specified work at the specified price, in contrast with '*estimate*' which frequently is no more than a general indication of roughly what the job is likely to cost. But such usage is by no means universal, and in each case one has to ask objectively whether it was a firm offer by the supplier or not.

4.6 DAMAGES AND OTHER REMEDIES

Earlier when considering 'General Principles of Contract' (section 4.3) we noted a distinction between terms and misrepresentations. We also saw that the reason for the traditional attempt to classify terms into conditions and warranties arises out of the desire to pin particular remedies to particular breaches of contract. So where one party breaks an express or implied term of the contract, the remedies of the other party depend on the importance of the term broken or the seriousness of the breach. Generally the innocent party seeks financial compensation for loss, although the circumstances may also entitle him to cancel the agreement.

4.6.1 Termination/Cancellation

Where the guilty party breaks a condition of the contract (for example, a seller delivering unmerchantable goods) or breaks a contract seriously, the innocent party may treat such a breach as a repudiation of the entire contract. This will discharge the contract, releasing him from his contractual obligations (for example, to pay the contract price). Businessmen often describe this as '*cancellation*'. of the agreement. It is a sanction which is seldom exercised, as most businessmen wish their relationships to continue, and prefer to use the right to terminate the agreement as a factor in the bargaining process to stimulate the other party into activity.

From the buyer's point of view in a sale of goods context this remedy is normally described as '*rejection*'. Thus where the seller is in breach of contract by delivering goods which do not match the specification, or are defective or unsuitable, the buyer may reject them by refusing to take delivery and, if he has paid money in advance, obtain a full refund. (This is why in a consumer context a shopper who has been sold faulty or misdescribed goods should obtain a complete refund, and need not be content with a credit note or even an exchange of goods.)

4.6.2 Acceptance

However, the right to reject goods for breach of condition is taken away in a sale of goods where the contract is not severable and the buyer has accepted the goods or some of them. (A 'severable contract' most frequently occurs where goods are to be delivered by installments and each installment is to be separately paid for.) The key factor here is *acceptance*. It can be express or implied; for example, where the buyer keeps the goods for some time without rejection (perhaps while he uses up existing stocks), or resells the goods and delivers them to the sub-buyer, or makes up raw materials. In all these cases, assuming that the buyer has had the chance to examine the goods, his conduct will amount to acceptance and he will be robbed of his right to reject the goods, even though he may not realise that the goods are defective or in some other way do not comply with the contract terms. This provision seems tough on the buyer since, although he may be complaining about a latent defect, which was not discovered until some time after delivery, he may not reject the goods. Yet this provision seems less severe when one bears in mind that the buyer's other remedy remains for breach of condition (that is, damages).

4.6.3 Damages

If the innocent party is content with compensation for his loss, it is immaterial whether he is complaining of a breach of condition or warranty, or a serious breach of an intermediate stipulation. He may recover *damages* for *any* breach of contract. The purpose of damages is to put the innocent party financially in the position in which he would have been if the contract had been properly performed—that is, compensation for his actual loss. The object is not to punish the guilty party. So if a breach of contract leaves the innocent party no worse off, only nominal damages will be awarded. Thus if a seller fails to deliver goods, and the buyer is able to buy similar goods from another supplier, at the same or a lower price than the contract price, the buyer will have suffered no loss and will be foolish to bring an action for breach of contract. Conversely, if a buyer wrongly refuses to take delivery, and the seller resells to a third party at a loss because the market price of the commodity has dropped, the buyer will be liable in damages for the difference between the contract price and the lower market price.

Yet a rider must be placed on the above remark. The damages awarded will cover only those losses which are not too 'remote', not every item of loss which flows from the breach and has a causal connection with it. As a matter of policy the contract-breaker is held responsible only for those losses which he could have had in mind at the contract moment as being likely to result from a breach of contract by him. Could he contemplate the loss as a real danger or as something liable to result? The rules on this 'remoteness of damage' topic stem from a nineteenth century case called *Hadley* v. *Baxendale*. The case shows that the contract-breaker may contemplate the loss in this way, either because it is normal and arises naturally in the ordinary course of events (rule 1) or because, though abnormal, it should have been anticipated in the light of special facts known to both parties at the contract moment (rule 2).

The examples given above illustrate the operation of rule 1. Clearly if a seller does not deliver the correct goods on time, the buyer may need to acquire them elsewhere; this is a normal consequence of the seller's failure, and the seller requires no special knowledge to realise that this will happen. Similarly, a buyer knows full well, when he agrees to buy goods, that if he refuses to take delivery, the seller will be left with the goods on his hands, and will be forced to sell them for the best price on the date when delivery should have taken place; and it is possible that the market price may have fallen (or risen) meanwhile.

Let us take an example from a capital project. Suppose that a contractor agrees to design, supply and install a pipeline for conveying flammable chemicals in a factory. The supplier breaks the contract either by carelessly installing the pipeline or by using unsuitable materials. The result is that the chemical leaks, an explosion occurs and the factory is burnt to the ground. The customer will be put to the expense of building a new factory. Until rebuilding is complete, production will be at a standstill with the consequent loss of profits. The contractor must compensate the customer for all these items, none of which is too remote.

Many suppliers consider that these rules making them liable for what is sometimes called 'consequential loss' are harsh, particularly as the total damages may far exceed the contract price for the installation, let alone the expected profit on that contract. On the other hand, why should the blameless customer stand the loss as compared with the contract-breaker? The problem is particularly acute when comparatively small sub-contractors are asked to tender for work connected with the petrochemical, nuclear power or offshore oil industries. Faced with the prospect of enormous losses resulting from comparatively small contracts, the contractor has a number of choices:

(1) he may decide that the risks are too great and refuse to tender for the job;
(2) he may seek insurance cover, although it is said that sometimes unlimited insurance cover is unobtainable, or only at exorbitant premiums;
(3) he may include an exemption clause in his conditions of sale, make absolutely sure that he wins the battle of the forms so that the clause becomes part of the particular contract, and hope that he will be able to prove that the clause is a reasonable one under the Unfair Contract Terms Act 1977 (see section 4.7).

In practice many suppliers adopt the latter two courses, if only because their insurers may insist on their using an exemption clause.

4.6.4   Exchange

There is no remedy in English law of *exchange*. So where a seller delivers unsuitable goods which the buyer rejects, the buyer cannot insist on the seller providing a substitute or replacement. As we have seen, the buyer's rights are to his money back and damages.

4.6.5   Specific Performance

The remedy of *specific performance* is rarely available in contracts for the supply of goods or services. It is an order of the court compelling the defendant to perform his part of the contract. It is a discretionary remedy which is not available where damages are an adequate remedy. Since in relation to goods or services, it is almost always possible to obtain another supplier at a price, compensation for the additional cost is an adequate remedy for the customer. Specific performance is generally ordered in contracts for the sale of land, where it is not enough to compensate the buyer for his loss of bargain, as he wants the particular piece of land and no other, and all plots of land are unique.

4.6.6   Misrepresentation

In commercial disputes undue emphasis is placed on *misrepresentation*, partly because its significance is misunderstood. Where the plaintiff (the person bringing the legal action) seeks only compensation, he need only prove that a breach of contract has occurred. This is generally no more difficult than proving that the contract was induced by a misrepresentation, and the recovery of damages for misrepresentation involves more complex matters too.

Nevertheless it is necessary briefly to explain the remedies for misrepresentation. It makes the contract '*voidable*'—that is, valid unless and until cancelled by the innocent party. If the contract is thus set aside, the innocent party should be restored to the position in which he was prior to the contract. Thus a buyer would have his money refunded. The remedy is usually described as '*rescission*'. It may be lost in a number of situations—for example, where the plaintiff has 'affirmed' the contract by treating it as continuing after learning of the misrepresentation. It follows that the plaintiff must seek this remedy quickly once the untruth of a statement has been discovered.

Originally damages could be awarded only where the misrepresentation was fraudulent, in that the misrepresenter knew that his statements were untrue, or was reckless in the sense that he did not care whether they were true or false. However, the Misrepresentation Act 1967 empowers the court to award damages even for innocent misrepresentation. The misrepresenter has a defence to this claim, if he can show that with good reason he believed what he said to be true. The effect is that suppliers who ought to know better in view of their expertise, when compared with their customers, will find it difficult to rid themselves of liability. This is sometimes described as 'negligent' misrepresentation in the sense that the defence is available to someone proving that he was not careless in making his statement.

4.6.7 Liquidated Damages and Penalties

It is not difficult to guess that many disputes turn on the extent of the liability of the contract-breaker. It may be plain that the contract has been broken, but the arguments may rage as to what items of loss should be included in the compensation, and how to quantify those items. In an attempt to short-circuit such arguments over remoteness and measure, it is common practice to include in major contracts what are variously called penalty clauses or liquidated damages clauses. Confusingly the expressions are used differently by lawyers and businessmen. We shall naturally use the expressions in their correct legal sense!

A *liquidated damages* clause is a clause where both parties endeavour to make a genuine pre-estimate of the loss to flow from a particular breach of contract, such as late delivery or late performance. This figure is included in the contract to fix the liability of the contract-breaker. If the contract is broken, the plaintiff is entitled to the stated sum as liquidated damages—no more, no less. It is irrelevant that his actual loss proves to be higher or lower; for if that were a relevant factor, the clause would have failed in its objective of obviating such arguments.

If, however, an inflated figure is put in by the customer to coerce the supplier into performing on time, it is a *penalty* and is void and irrecoverable. This does not mean that the innocent party is unable to recover damages. The parties are left in the same position as if the clause had been omitted and will need to negotiate the amount of liability.

We have seen before that labelling clauses with particular titles or phrases does not make the problem go away. So here describing a clause as a 'penalty' clause or 'liquidated damages' clause does not resolve the problem. The question will still be whether or not the figures represent a genuine attempt to estimate the ultimate losses, on the facts known to the parties at the time the contract is entered into.

A further complication arises from the practice of describing some exemption clauses as 'liquidated damages' clauses. For example, in many construction contracts a clause is included limiting the contractor's liability to, say, 0.5 per cent of the contract price per week up to a maximum of 5 or 10 per cent of the contract price. This is probably neither a penalty clause nor a liquidated damages clause. Bearing in mind that damages may far exceed the contract price, a clause limiting the liability of the contractor to a small percentage of the contract price is in reality an exemption clause and subject to control by the Unfair Contract Terms Act 1977 to which we now turn.

4.7 EXEMPTION CLAUSES— THE UNFAIR CONTRACT TERMS ACT 1977

4.7.1 Introduction

Generally the courts and Parliament in recent decades have adopted a hostile attitude to exemption clauses. The arguments for this policy can be simply stated. Suppliers are not compelled to enter into contracts. If they choose to do so, they should expect some responsibilities, to counterbalance the price to be charged which is freely negotiated. So it is not unreasonable to expect them to supply equipment and services of the right type and quality, and at the correct date in return for their charges. If the supplier does not fully comply with his contractual obligations, why should he expect the customer to comply with his by paying the price in full? The supplier's reply is to point out that a single breach of contract by him may cause enormous losses to the customer, as we saw when considering damages, and that it is fair for the supplier to guard against such claims, which could well drive him out of business, by incorporating protective clauses.

At common law, various tactics have been developed by the courts to defeat exemption clauses. For example, if the clause has not become part of the particular contract in dispute, because the supplier lost the battle of the forms or attempted unsuccessfully to introduce the clause in a post-contractual document such as an invoice, then clearly it is of no avail. However, since the legislative attack on exemption clauses in the 1970s, culminating in the Unfair Contract Terms Act 1977 ('the UCTA'), customers have tended to look first to the Act as a weapon.

Exemption clauses, or '*disclaimers*' as they are sometimes called, may be put into two categories:

(1) exclusion clauses, excluding all liability;
(2) limitation clauses, accepting some liability but reducing it to a stated sum, or to the contract price or a percentage of it.

For the purpose of this classification, guarantees and similar clauses may be put in the second category, since normally the manufacturer agrees to repair or replace defective parts free of charge for a stated period after delivery (usually 12 months), but excludes all further liability, notably for consequential loss. The distinction between exclusion and limitation clauses is important, as the courts view exclusion clauses with greater disfavour, on the basis that any supplier has sufficient resources to be able to stand some financial responsibility for the error of his ways.

Two general points must be made on the 1977 Act before considering its provisions in more detail. First, it does not impose any liability on suppliers; however, if it can be shown that the supplier (1) is *prima facie* liable because of a breach of one of the statutory implied terms in the 1979 or 1982 Acts or for some other reason and (2) is protected against that liability by an exemption clause, only then need the customer bring the 1977 Act into play. Secondly, in spite of its misleading title the UCTA cannot be used to attack any term simply because it seems to be 'unfair'; the UCTA is concerned only with exemption clauses.

As a sidenote, we must mention that some of the provisions in the Act draw a distinction between '*consumer dealings*' and other transactions, and go so far as to ban exemption clauses where businessmen supply ordinary consumer goods to private customers. However, in this book a discussion of these aspects of the Act would be inappropriate, and we shall confine our attention to transactions between one business and another. In such business transactions the basic test to be applied to exemption clauses can be stated concisely. Can the supplier prove that the exemption clause is reasonable? If not, it has no effect on the supplier's liability.

The most important provisions of the UCTA apply to three main sets of circumstances.

(1) *Supplies of goods.* If the supplier attempts to exclude or limit liability for breach of the implied terms relating to description, merchantable quality and fitness for purpose, the clause must satisfy the reasonableness test. So the Act controls typical clauses in conditions of sale, whereby the supplier attempts to evade liability for supplying goods which do not match the specification, are defective or fail to meet the customer's specific requirements.

(2) *Negligence.* The reasonableness test is also applied to a clause, where the supplier attempts to exclude liability for breach of any obligation to take reasonable care or to exercise reasonable skill; for instance, where the supplier of a service such as a structural engineer gives careless advice, or the supplier in a contract for work and materials carelessly installs plant. (If death or injury results, the clause is void.)

(3) *Breach of contract.* A clause is also subjected to the reasonableness test where the defendant attempts to cut out or reduce liability for any breach of contract. This part of the Act, however, applies only where the contract is made on 'written, standard terms of business'—that is, not to tailor-made, individually negotiated contracts. A typical example of a clause falling within this provision relates to attempts by suppliers to avoid responsibility for late delivery or non-delivery. Technically a *force majeure* clause will be covered by this provision, but there is little risk of being unable to prove such a clause to be reasonable.

4.7.2 The Reasonableness Test

In all the above situations the supplier must prove that 'the term satisfies

the requirement of reasonableness'. This requirement is set out in the
'*reasonableness test*' contained in the Act: 'The term shall have been a fair
and reasonable one to be included having regard to the circumstances which
were, or ought reasonably to have been, known to or in the contemplation
of the parties when the contract was made'. This test does not obviously
take us much further than the original expression. When the Act was drafted
it was recognised that the test was somewhat vague, and so to enable it to
be applied more easily a set of five 'guidelines' was included. Technically
these helpful guidelines apply only in relation to the implied terms in
contracts for the supply of goods (category (1) above) but in practice the
courts are applying similar criteria generally.

The three most important guidelines require the following questions
to be asked. Was the customer of equal bargaining power with the supplier?
Did the customer have a choice as to contract terms? Did the customer
have notice of the exemption clause? If the answer to these questions is
'Yes', the customer will have great difficulty in upsetting the exemption
clause.

In slightly more detail, and listing them in the same way as the Act,
the guidelines are as follows:

(a) 'The strength of the bargaining positions of the parties relative to each
other, taking into account (among other things) alternative means by
which the customer's requirements could have been met.' So a clause
may be reasonable when dealing with a large customer, such as an oil
company well able to look after itself commercially, but unreasonable
when dealing with a small customer with little commercial influence.
It can be seen that another factor is whether the supplier is in a
monopolistic position, or whether the customer had a wide range of
choice in choosing the supplier.

(b) The second guideline asks 'whether the customer received an induce-
ment to agree to the term'. For example, some suppliers adopt a 'two-tier'
system by giving the customer a choice of terms, with a higher price
and no exemption clause or a lower price with an exemption clause, a
system likely to be reasonable. This guideline also asks whether the
customer 'had an opportunity of entering into a similar contract with
other persons but without having to accept similar terms'. For example,
one supplier may insist on an exemption clause, but another supplier
with a suitable product may be prepared to take full responsibility. If
the customer chooses to contract with the first supplier on disadvan-
tageous terms, that is his own funeral.

(c) 'Whether the customer knew or ought reasonably to have known of the
existence and extent of the term.' Although it is common commercial
practice to hide away exemption clauses in the small print on the back
of quotations—often printed in pale green on white paper, or grey on

yellow or something similarly illegible—this practice is likely to make the clauses unreasonable. A bold clause on the front of the quotation is more likely to be effective. The court will take into account the custom of the trade, and any previous dealing between the parties in deciding whether or not the customer knew or should have known of the clause.

To sum up, the supplier will be in the safest position where the customer is a large commercial concern, could obtain the goods or services elsewhere but chooses to contract with the supplier on prejudicial terms, and enters into the contract with his eyes open. Conversely a small customer given 'Hobson's choice' and unaware of the exemption clause is most likely to be unaffected by an exemption clause.

4.8 PRODUCT LIABILITY

In this chapter we have concentrated on the contractual position between the contractor and the employer. Although it would be inappropriate to attempt to explain the law of *negligence* in detail, even as applied to the relationship between a manufacturer and an end user, we must at least mention this area of law. For more than 50 years manufacturers have been liable to end users, whether or not there is a contractual relationship between them, if they launch into the market place products which are unsafe, dangerous or likely to cause personal injury or damage to property, but only where defects in goods result from their negligence. In other words, manufacturers have not traditionally incurred strict liability for their products, except to their immediate customers by virtue of the agreement between them. The legal position changed in March 1988 when the Consumer Protection Act 1987 came into force. Now manufacturers and importers are strictly liable for dangerous products but only to private users. So as between one business and another the position will remain the same. The existing law of contract and negligence continues side by side with the new law.

4.9 CONCLUSIONS

Since a project is essentially a one-off activity, it is quite likely that an organisation will not have sufficient resources to run a project entirely from its own internal resources. This can be simply a question of available manpower, or it can be caused by a lack of the necessary skills. The latter situation is exacerbated by the range and degree of specialisations in modern technology. Consequently the use of contractors is likely to increase, especially in high technology applications. For these reasons, the selection of contractors has been discussed in this chapter. To contrast with these, some notes have also been added that deal with tendering: project engineers are equally likely to be involved with tendering.

Contracts are essentially of two types, and these are fixed price or cost plus contracts. Each of these has been described and discussed, along with the variations that occur between these two extremes. In general, a fixed price contract is used when a requirement is accurately defined, since it offers a customer the best value for money and is comparatively easy to control. In contrast, the cost plus form of contact is used when the requirement is ill-defined. By its nature it can be difficult to control, and be expensive for the customer. However, there are some activities, such as development work, which no contractor would be likely to tender for on a fixed price basis.

Contracts do not just involve the purchase of services, but they also include the purchase of goods, and consequently contract law has a very wide ranging significance. After a description of the general principles of contract law the duties of suppliers have been discussed here. This is firstly for the supply of goods and secondly for the supply of services. Since not all contracts run smoothly, it is as well to know the possible legal implications. The possible outcomes have been presented in section 4.6, and these can take the form of damages, or other remedies such as termination of the contract.

An important reason for appreciating the legal position, or obtaining legal advice, is that prevention is better than cure. In other words, it is much the best policy to avoid litigation, since otherwise it may only be the lawyers who benefit. The position with international contracts is much more complex; while there are international lawyers there is no international law as such. International lawyers have to decide which legislation is applicable from which country, and then arrive at a mutually acceptable settlement.

4.10 DISCUSSION QUESTIONS

1. Why is there often a need to employ contractors in project work? What criteria can be used for selecting a contractor?
2. Describe the different forms of contract, and identify their strengths and weaknesses. Give examples of project work that justify the use of each type of contract.
3. How can a contractor improve his chances of winning contracts, yet still achieve profitability?
4. What actions should a contractor take to ensure satisfactory financial arrangements in:

   (a) overseas contracts?
   (b) domestic contracts with a long timescale?

(c) contracts where there may be disputes?

5. Discuss how the processes leading to a contract (offer, counter offer(s), acceptance) affect the terms of business on which the contract is made.
6. Distinguish between *terms* and *representations*. Discuss the types of term and the different forms of misrepresentation.
7. What is understood by *privity of contract*, and *agency* (with particular regard to apparent authority).
8. Discuss the difference between *force majeure* and *frustration*.
9. For the sale of goods, the suppliers' duties are covered by the 1979 Sale of Goods Act, and one particular area that is covered is quality and fitness. What is meant by: *merchantable quality* and *fitness for purpose*? What are the limitations of guarantees?
10. When a service is being supplied, explain how the 1982 Supply of Goods and Services Act ensures that the time taken and charge for the service are reasonable. How is the extent of liability limited?
11. If a contract has not been executed according to the agreed terms and conditions, what form can the damages and other remedies take?
12. How can a supplier show that an exclusion term is reasonable?

Chapter 5
Safety and Risk

5.1 INTRODUCTION

The requirements for safety are ever increasing, yet the difficulties in ensuring adequate safety in any operation are also increasing for a variety of reasons.

The growth of technology has increased the size and complexity of industrial processes, as well as increasing the range of materials involved. Furthermore, the widespread use of continuous, larger-capacity processes can lead to increased quantities of materials being held in storage. Often these materials have to be stored at high pressure and/or extremes of temperature. The consequences of any accident in a large plant can be much more serious, yet the risks are also more difficult to determine. At the same time the safety requirements have been improving throughout the world; these effects all combine to increase the challenge to engineers.

The evolution of safety legislation in the UK is described by Harvey (1983), who points out that as late as the 1850s boiler explosions were regarded as acts of God, and were described as such on death certificates. In Europe, legislation in recent years has developed along parallel lines in Germany, Norway, Sweden and the UK. British legislation often provides a lead for past and present members of the Commonwealth, and the British and German membership of the EEC is likely to influence future EEC legislation. In the USA the Occupational Safety and Health Administration (OSHA) has been monitoring European legislation, and is thus likely to be influenced by it. Consequently, the material in this chapter is based on British legislation, the Health and Safety at Work Act (1974). Since legislation is not uniform worldwide, it is quite possible for a machine designed and used in one country to contravene the safety requirements of another country.

Engineers, of course, are in a key position to ensure safety, since they are involved with the design, development and manufacture of equipment. Safety should be considered as early as possible in each stage of a project, and it should always be treated as a challenge, not as a chore. Engineers are also likely to be the only people who are able to make a sensible estimate of the risk of an accident occurring and, if money is to be spent wisely on improved safety, then risk has to be evaluated.

In coping with the problems of risk, which are inherent in any engineering operation, the project engineer will need to have some understanding of the legal issues relating to health and safety at work. The law itself is extensive and detailed; this chapter will not give all the answers. But, by discussing some of the principal features of the law and its administration, it is hoped that project engineers will be guided to ask the right questions of those particularly able to help.

This chapter ends with a discussion of risk, the attitudes towards risk, and how safety can be ensured.

5.2 TECHNICAL LEGAL MATTERS

Before turning to the substance of Health and Safety law, it is necessary to get two technical legal matters out of the way: first, what are the sources of the law? secondly, in which courts are they enforced?

5.2.1 Sources of Law

Statutes

Under the English legal system, it needs to be appreciated that there are two distinct sources of law: legislation (including regulations) and case law. The main source of law today is *legislation*—that is to say, *statutes* or Acts of Parliament which started off as Bills, usually drafted by civil servants acting under instructions from their political masters, the Ministers, and which have then passed through various stages of debate in Parliament, until finally they are passed into law when they receive the Royal Assent.

Because of the complexity of the issues relating to health and safety and the need for flexibility, the detail of such legislation is 'fleshed out' in *regulations*, made under the authority of Acts of Parliament. Such regulations (technically known as 'Statutory Instruments') do need to be presented to Parliament, but only very rarely are they the subject of debate in Parliament. Thus they will come into effect without the need for protracted debate and discussion. These regulations are sometimes described, generally, as '*secondary legislation*', having been made according to rules contained in '*primary legislation*'—that is, the Acts of Parliament themselves.

In addition to Acts and Regulations which are made in Britain, the institutions of the European Economic Community may also make law relating to matters of health and safety. Its directives are usually brought, formally, into force in Britain by being incorporated in regulations (see above); for example the Control of Health at Work Regulations 1980, and the Safety Signs Regulations 1980 are both examples of British law incorporating directives laid down by the EEC.

Case Law

Besides legislation there is also case law—rules of law laid down in cases decided in the most authoritative of our courts (the High Court, the Court of Appeal and the House of Lords), which are then binding on the lower courts—the Crown Court and the County Court. Case law is of two distinct types. The first involves the *interpretation of statutes*. Although the statute law mentioned above is often of breathtaking complexity, it is inevitable that the language of a statute will on occasion be ambiguous, or will not apply clearly to a given set of facts. The courts may, therefore, be called upon to interpret or rule upon the meaning of particular statutory provisions.

The second type of case law is the so-called *common law*. These are principles of law that have been developed in cases, decided over the centuries by the judges of the higher Courts, where statutory rules are *not* involved. (One example of a set of common law rules relates to the formalities that must be gone through in order to enter a legally binding contract, as discussed in section 4.3; another familiar example is found in the principles of the law of negligence, whereby if one person acts in breach of a duty of care towards another (known in law as his 'neighbour'), the person in breach of the duty may be liable to pay compensation for any damages suffered by the injured party.)

As will be seen, Health and Safety law is to be found both in the rules of statute law and the principles of common law.

Codes of Guidance and other Official Advice

Under the British constitution, rules having the force of law must either have been through the appropriate Parliamentary process (statute law) or been handed down by the courts (case law). In the area of health and safety, however (and increasingly in other areas as well), rules of practice are to be found in a variety of *Codes of Guidance* or *Guidance Notes*. Although these are not law in the technical sense, they frequently have a 'quasi-legal' status, and are of great practical importance; for example, they may well be taken notice of in a case that comes to court (rather like the Highway code may be taken note of in a Road Traffic case).

5.2.2 Courts of Law (and Tribunals)

The Court of Law (or Tribunal) which deals with questions of health and safety varies, depending on the nature of the dispute which is being taken to court.

The Criminal Courts

If the issue involves a *prosecution*, by one of the health and safety enforcement agencies or the police, for a criminal offence that would result in some kind of penalty (such as a fine or even imprisonment), the case will usually be dealt with in the *magistrates'* court or, for more serious offences, the *Crown* Court.

The Civil Courts

Where the case involves a civil matter, say an action for breach of contract or breach of statutory duty, the case will be heard in the *County Court* or, where the sums of money in dispute are larger, in the *High Court.*

Appeals

Appeals from either the criminal courts or the civil courts will be dealt with usually by the *Court of Appeal*; in very rare instances, there may be the possibility of a second tier of appeal to the *House of Lords* (sitting in its judicial capacity).

Tribunals

The *Industrial Tribunals* and, on appeal, the *Employment Appeal Tribunal* have a specially defined jurisdiction in health and safety cases, to deal with particular types of issue, notably matters relating to prohibition and improvement notices (see section 5.5.5).

5.3 THE BACKGROUND TO THE LEGAL STRUCTURE

Understanding the details of the law will be assisted by appreciating, in general terms, the *objectives* of the modern law on health and safety, and being aware of certain basic aspects of the *history* of the law.

5.3.1 The Objectives of the Law

There are four clearly identifiable objectives to health and safety law: punishment, compensation, prevention and participation.

In general, the law tends to be rather better at picking up the pieces when things have gone wrong, as opposed to preventing undesirable events occurring in the first place. Thus, much health and safety law has been developed in order to provide mechanisms for imposing punishments if things go wrong, especially against employers who neglect to provide a safe working environment. These will be imposed in the criminal courts (section 5.2.2).

In addition, health and safety law offers a basis for awarding *compensation* (damages) for those injured or killed in work-related accidents. These will be decided by the civil courts (section 5.2.2).

But injuries and deaths are, above all, to be prevented. Thus much of the modern law aims at the *prevention* of accidents, for example, by prohibition and enforcement notices.

Underpinning all the legal structure, however, and perhaps most important of all, is the notion of *participation*. All sides in industry, employers, workers and government, are involved—in the legal structure, in investigating accidents, researching accident prevention, enforcing the law when things go wrong and so on.

5.3.2 The History of the Law

The backbone of the modern law is now to be found in a statute, the Health and Safety at Work Act, 1974. In this section, we will look at a number of the main events leading to the passing of that Act by Parliament.

There has been legislative intervention by Parliament to protect particular classes of workers in particular industries for many years. For example, as long ago as 1802, the Health and Morals of Apprentices Act was passed by Parliament, an Act designed, as the name suggests, to regulate the conditions under which children could be employed. The first major piece of health and safety legislation was Lord Shaftesbury's Factory Act, 1833. This set the pattern for subsequent legislation, in that a variety of Acts was passed over the years, relating to specific work contexts. Thus, by 1970, there existed the Mines and Quarries Act, 1954; the Factories Act 1961; and the Offices, Shops and Railway Premises Act 1963; together with voluminous secondary legislation made under these Acts.

There was also steady progress with the common law as well. For example, in an important case in 1840, a girl injured in an accident at a mill successfully sued her employer for breach of his duty of care towards here. In its day this ruling represented an important development in the law of negligence. And in an even more important case in 1898, *Groves* v. *Lord Wimborne*, it was held that employees who had suffered an injury could sue their employers for compensation where those employers were found to be in breach of their statutory duties under legislation, such as the Acts mentioned above. (The importance of this case was that, under the Acts themselves, the State imposed defined penalties by way of rules of criminal law; but, of course, this did not compensate an injured worker for his injuries. The case opened up the means of suing for damages for injuries incurred where the employer was in breach of his statutory duty; this could be done without the need for proving negligence, in its common law sense.) In this way, the statute law and the common law became inextricably intertwined.

5.3.3 The Robens Committee

It was into the maze of common law and statute law that the Robens Committee plunged in 1970 when an official review of health and safety law was set up by the government. The Committee was extremely critical, finding, among other points, that:

(1) the law was not effective; there were thousands of work-related deaths and injuries each year;
(2) the law was complex;
(3) enforcement of the law was very confused;
(4) there was widespread apathy in industry.

Its report, published in 1972, sought to change all this. The result was the enactment of the Health and Safety at Work Act 1974. Compared with the fate of the recommendations of many official committees, this was a surprisingly rapid result. The way this legislation has been brought into effect has not, as we shall see below, resulted in as clear-cut a statement of health and safety law as might have been hoped for. Earlier legislation, such as the Factories Act 1961, the Mines and Quarries Act 1954 and the Offices, Shops and Railway Premises Act 1963 all remain in force (even though the long-term goal is to repeal them); the principles of common law still supplement the statutory rules; the methods of compensation for work-related accidents remain outside the legislative framework (and are found instead in the Common law and social security law). Nonetheless, the 1974 Act does represent three major advances on what existed before:

First, the Health and Safety at Work Act applies to *all* employed workers, not just those working in particular work environments such as factories, mines, offices and shops etc. Indeed, it was estimated that some 8 million workers obtained legal proection for the first time when the Act was passed.

Secondly, there is now a basic national enforcement agency to enforce health and safety law—the *Health and Safety Executive*—instead of a range of enforcement agencies set up under each individual Act of Parliament. (Local authorities also retain some enforcement powers.)

Thirdly, there is now a national body able to engage in research and the publication of advice to industry: the *Health and Safety Commission.*

5.4 THE MODERN INSTITUTIONAL FRAMEWORK

There is a great variety of institutions responsible for the development and creation of health and safety law.

The Government

Primary responsibility rests with the Government and Civil Service, in particular the Secretary of State for Employment and the Department of Employment. Other departments, such as the Department of the Environment and the Department of Energy, also have an interest.

The Health and Safety Commission

This is a 'quango' established by the government under the Act of 1974 to secure the health and safety of those at work, by looking after the administration of health and safety law. It has a wide range of powers to enable it to carry out its tasks. One of the most important is to order the establishment of investigations or enquiries after an accident has occurred. The Commission is also responsible for developing the law—for example, by the preparation of codes of practice and drafting new regulations where needed (for approval by Ministers and Parliament); it also undertakes research.

The Commission has a chairman plus up to nine other members, and it is supported by a secretariat. Furthermore it is advised by a member of specialist advisory committees on, for example, Major Hazards, Toxic Substances, Dangerous Substances, Dangerous Pathogens and Nuclear Safety.

The Health and Safety Executive

The Health and Safety Executive is responsible for the enforcement of health and safety legislation. The structure of the Executive is complex to describe, though an attempt is made in figure 5.1. It operates from central offices in London and Bootle, together with 21 area offices throughout the country, each controlled by an area director.

The Executive is organised into a number of branches comprised, in the main, of the separate inspectorates that existed prior to the passing of the 1974 Act. These include: the Factory Inspectorate, the Explosives Inspectorate, the Nuclear Installations Inspectorate, the Mines and Quarries Inspectorate, the Alkali and Clean Air Inspectorate, and the Agricultural Health and Safety Inspectorate.

Within each area office there is also located a *national industry group* designed to concentrate attention on the problems of particular industries—for example, construction, docks, electricity, engineering, foundries, shipbuilding, steel, textiles, and wire and rope making.

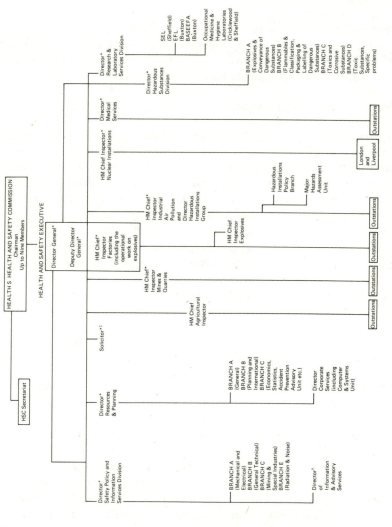

Figure 5.1 Organisation of the Health and Safety Commission and Executive [Reproduced with the permission of the Health and Safety Executive]

In exercising its powers of enforcement, the Health and Safety Executive is, in effect, acting as the agent for the Health and Safety Commission. However, the Executive has sole discretion as to what methods of enforcement should be adopted. Although much of the law is drafted in terms of criminal offences, the Executive in fact operates a wide measure of discretion in its enforcement policy, and relies more on persuasion and its powers to issue notices than on prosecution as the primary means of enforcing the law.

Local Authorities

Despite the primary purpose of the Health and Safety Executive to enforce safety legislation, local authorities retain some power of enforcement, particularly in relation to retail establishments. Local authorities also check on the safety in offices operated by Central Government Departments. Full details are in the Health and Safety (Enforcing Authority) Regulations, 1977.

Safety Representatives and Safety Committees

In work places with a trade union which has been recognised by the employer for the purposes of collective bargaining, the trade union may appoint a *safety representative* who represents the employees, and who must be consulted by the management. The safety representative has powers to investigate potential hazards and dangerous occurrences. He may inspect premises (every 3 months if needs be); he is entitled to time off work for his work as a safety representative and to attend training offered by his union or the TUC. The number of safety representatives per place of work, will obviously depend on the size of the work place and work force. Where two or more safety representatives so request, a *safety committee* must also be set up. (Safety representatives and safety committees which are set up under the Act of 1974, and have powers granted under the Act, are to be distinguished from non-statutory *safety officers*, who may be appointed as a consequence of enlightened management practice, but who have no legal powers attached to their appointment.)

5.5 THE MODERN LAW

5.5.1 Preliminary Points

Having looked at the variety of bodies with responsibility for administering, developing and enforcing health and safety law, we now turn to the substance of the law which they have to enforce. In considering the legal rules, some preliminary points need to be borne in mind.

(1) Although the bulk of the statutory duties relating to health and safety at work is placed on *employers* (that is, the management), there are some duties which are also placed on *employees*. Safety at work is everyone's responsibility, and this is recognised in the law.

(2) The law does recognise that health and safety laws have to operate in the real world of business enterprise. Standards of safety cannot be made so high that industry becomes completely uncompetitive. Some duties are so important that they are made *absolute* duties—that is, no exceptions are allowed. Most duties are, however, *qualified* to some extent.

The law has a number of 'code' words of phrases to achieve this objective. If the words 'shall' or 'shall not' are used (for example, 'every dangerous part of any machinery *shall* be securely fenced'), this imposes an absolute standard on the employer. If, in the example given, a machine cannot in fact be securely fenced, then, in law, the machine cannot be used; the duty to fence is an absolute duty. The importance of guarding and the means of achieving the necessary performance are discussed further in section 5.9.

On the other hand, if the phrases 'so far as is practicable' or 'best practicable means' or 'reasonably practicable' are used, these imply a high standard of care, but not an absolute one; it is a *qualified* standard. There is deliberately implicit in such a standard, scope for discussion and negotiation as to what *is* practicable. Such standards inevitably have 'fuzzy' edges, and thus the law is on occasion criticised for being vague; but is it not sensible to allow *some* flexibility in the law and its application?

As a guide, where the law requires steps which are *practicable* (this means that if it is feasible to do something about a danger which is known to exist), then such action must be taken whatever the cost. If the requirement is to do what is *reasonably practicable*, then it is possible to balance the nature of the danger or risk that may occur, against the cost and inconvenience of doing something about it.

5.5.2   The Health and Safety at Work Act 1974

The duties under the Health and Safety at Work Act may be classified under three headings: (a) duties imposed on employers towards their employees; (b) duties imposed on employees and (c) duties imposed on others. Each category will be considered in turn. At the end of this section, the powers of the inspectors from the Health and Safety Executive will be considered.

5.5.3 Duties of Employers towards their Employees

Over-riding General Duty

The over-riding duty created by the Act of 1974 is expressed thus:

'It shall be the duty of every employer to ensure, so far as is reasonably practicable, the health, safety and welfare at work of all his employees.'

This is the fundamental proposition of law, which is then amplified in a huge variety of ways in other more specific contexts. As will be noted, this provision incorporates the 'reasonably practicable' test mentioned above.

Perhaps surprisingly, the phrase 'health, safety and welfare' is not defined in the Act. It is clear that 'health' includes both mental as well as physical health; 'safety' is thought to refer to the absence of foreseeable injury; 'welfare' is a broader concept. 'At work' makes it clear that the duty extends to employees only while they are at work, and acting in the course of their employment.

Specific Duties towards Employees

Notwithstanding the generality of the above, the Act goes on to detail a number of examples of specific duties owed to employees. They relate to:

(1) the provision and maintenance of plant and systems of work;
(2) the use, handling, storage and transport of articles and substances;
(3) the provision of information, instruction, supervision and training relating to health, safety and welfare;
(4) the maintenance of the place of work itself and all means of access;
(5) the provision of a safe working environment.

All of the above are at the 'so far as is reasonably practicable' level.

Furthermore, all employers with more than five employees must adopt and publish a *safety policy*, which must be intelligible to employees, and also be regularly reviewed.

5.5.4 Duties Imposed on Employees

The Act also imposes a number of important duties on employees. Firstly, they have a duty to take reasonable care for the health and safety of themselves *and of others* who may be affected by their acts or omissions at work. Thus, refusal to use safety equipment or to fail to use safety precautions could render an employee liable to prosecution by inspectors from the Health and Safety Executive. A prosecution could also be started against those who behaved carelessly or negligently (such as by skylarking).

Secondly, employees are under a duty to co-operate with their employer in matters of health and safety. Refusal to do so can, again, render an employee liable to prosecution.

Thirdly, there is a duty imposed on employees not to intentionally or recklessly interfere with anything provided for the health, safety or welfare of employees. Breach of this duty may also lead to prosecution by the Health and Safety Executive.

5.5.5 Duties Imposed on Others

The Health and Safety at Work Act also imposes duties that extend beyond the direct employer/employee relationship.

The Self-employed

Where a person who is self-employed is working in a context where he comes into contact with others (who are not employees of his), he is under a general duty to conduct his undertaking in such a way as not to expose others to risk in health or safety.

Employers Generally

Similarly, employers are under a duty not to expose persons (who are not employees) to a risk in either health or safety (for example, holiday-makers staying on a farm, or visitors to a construction site).

Controllers of Premises

A controller of premises (such as a landlord) is also under a duty to keep premises, the means of access and exit, and any plant or substance therein or provided for use therein, safe and without risk to health.

Designers, Manufacturers, Importers and Suppliers

A range of duties is also imposed on the designers, manufacturers, importers and suppliers of 'any article for use at work' or 'substances for use at work'. The duties are: (a) to ensure that articles or substances are safe; (b) to carry out any necessary testing of the article or substance; (c) to provide adequate information about the article or substance; and (d) to do research to eliminate or minimise risks to health and safety. Special duties are also imposed on erectors and installers of plant and other articles and substances for use at work.

5.5.6   Powers of Enforcement

Enforcing the law sketched out above is, as already noted, the responsibility of the Inspectors of the Health and Safety Executive. To assist them in carrying out their tasks, they have been granted wide *powers of investigation.* Thus they have powers, among others, to enter premises, with or without the police; they may take their own equipment into a work place; they may take photographs of the work place; they may require the production of a company's books. Naturally, the information obtained might well be commercially extremely valuable. Thus, in general, they are forbidden to disclose any information obtained, save where legal proceedings occur.

Breach of any of the duties outlined above can, in theory, lead to a criminal prosecution by inspectors of the Health and Safety Executive. Furthermore, where there has been a major accident, or someone is reluctant to comply with the persuasive advice of an inspector, a prosecution may actually ensue.

But, as noted above, the inspectors are generally more anxious to prevent accidents occurring in the first place. To assist them in their preventative work, the inspectors have important powers to issue *notices.*

Improvement Notices

These may be issued where the inspector thinks a breach of the law is occurring, or has occurred and is likely to occur again. The notice must state what specific law is, in the inspector's opinion, being broken; the reasons must also be given. The notice will require the breach of the law to be stopped within a period defined in the notice (at least 21 days). It is not necessary for the notice to prescribe how the breach is to be remedied, but this may be done.

Prohibition Notices

These are clearly more serious than an improvement notice and may be issued where an inspector is of the opinion that there is a 'serious risk of personal injury'. As with the improvement notice, the prohibition notice has to contain specified written details: what the serious risk of injury is; the relevant statutory provision being infringed (if any); and a directive that the relevant activity must no longer be continued. If the risk of injury is current or imminent, the prohibition notice may have immediate effect; if not, the operation of the notice may be deferred until a date specified in the notice.

Appeals

If a person on whom a notice is served objects, he may *appeal* to the

Industrial Tribunal within 21 days of the service of the notice. If this occurs, an improvement notice will automatically be suspended until the Tribunal has heard the appeal; in the case of a prohibition notice, however, it will be suspended only if the Tribunal agrees.

Even if no formal appeal is made, a notice may be withdrawn, if a person on whom it has been served satisfies the inspector either that the terms of the notice have been complied with, or that there was no good ground for serving the notice. Conversely, where the breach of health and safety law or other serious risk of injury is still present, a notice may be extended.

5.6   OTHER STATUTORY PROVISIONS

As mentioned at the outset of this chapter, although the long-term intention is to bring all health and safety legislation into a single legislative code, the present position, over a decade after the Robens Act was passed, is that other legislation (applicable in particular contexts) is still in force. It is not practical to go through these additional legislative measures in detail. Instead, a number of the most important features of that law will be highlighted.

5.6.1   The Factories Act 1961

Perhaps the most important additional legislation for the project engineer is the Factories Act 1961. As its name implies, this Act applies in factories. The term 'factory' is very broadly defined as a place where someone is employed in manual labour, making, altering, repairing or cleaning any article, and where the work is carried on as part of a trade for profit.

The most important duties under the Factories Act relate to the following matters:

(1) *Health and Welfare.* It is the duty of the factory occupier (usually, of course, the employer) to keep the premises clean and free from accumulations of dirt or refuse. In addition, there are rules relating to the adequate circulation of fresh air, sufficient heat and the provision of basic amenities such as lighting and toilets. Drinking and washing water are also to be provided.

(2) *Safety.* Important aspects of this legislation deal with the *fencing* of machinery in factories. (Indeed, it is under these provisions that prosecutions for offences are most frequently brought.) It is in this area that *absolute* duties (see above) are to be found; guarding is discussed further in section 5.9. Regulations made under the basic statutory provisions relate, for example, to Abrasive Wheels (1970) and Power

Presses (1965). It is also under the Factories Act 1961 that the Protection of Eyes Regulations were made, requiring the use of goggles or shields for certain defined operations (for example, blasting concrete or striking masonry nails).

In addition, there are special rules relating to *lifting gear*, such as hoists, lifts, cranes and ropes. These must be examined regularly and details contained in a written register. Again, this is an example of an absolute duty. Similar rules relate to steam and compressed air apparatus.

Enforcement of the Factories Act is now carried out by Inspectors from the Health and Safety Executive, using their powers under the 1974 Act.

5.6.2 Fire Precautions Act 1971

This brings together a number of provisions relating to fire precautions formerly found in different items of legislation. Enforcement here is by fire authorities rather than the Health and Safety Executive (except in cases where there is a special risk of fire or explosion, such as in an explosives factory, where a fire certificate will be issued by the Health and Safety Executive).

5.6.3 Other Legislation and Quasi-legal Standards

Detailed legislation, regulations and guidance are also to be found in relation to the matters listed in figure 5.2. Enforcement of all this legislation is also by Inspectors of the Health and Safety Executive, using their powers under the 1974 Act.

MAJOR SAFETY LEGISLATION
(in addition to the Factories Act)

Provision of Fire Certificate
(includes the specification of fire exits,
gangways and doors which must not be changed
without consultation)

Electricity Regulations
Kiers
Building Regulations
Reportable Dangerous Occurrences
Ionising Radiation
Power Presses
Abrasive Wheels
Highly Flammable Liquids and LPG
Eye Protection
Pollution (Air, Land and Water)
Electrical Installations in Flammable Atmospheres
Asbestos
Work in confined spaces
Handling of Food

QUASI-LEGAL STANDARDS AND GUIDANCE DATA

British Standard Specifications and Codes of Practice
Institution of Electrical Engineers Wiring Regulations
Noise
Lighting
Fire Prevention (FPA and FOC)
HSE Guidance Notes and Booklets on a wide range of
specific hazards: Toxicity of Materials, Safety in
Welding, Specific Process and Machine Hazards etc.
Codes of Practice developed by specific industries,
for example Rubber, Textile, Printing, Paper, Power
Presses, British Gas etc.

Figure 5.2   Additional sources of safety legislation and guidance

5.7 THE COMMON LAW

As mentioned at the start of this chapter, health and safety law is the product both of statute law and of the common law. There are two main headings under the common law which are relevant to health and safety matters: negligence and breach of statutory duty.

5.7.1 Negligence

To establish a claim for *negligence*, a plaintiff (here, the injured worker) must show three things:

(1) the defendant owed the plaintiff a duty of care;
(2) the defendant was in breach of that duty;
(3) as a consequence the defendant was injured.

In the present context, it has been held that an employer is under a general duty to provide a safe working environment for his employees by taking reasonable care to provide proper appliances, and to maintain them in a proper condition. In general, the duty extends only to protecting the employee against personal injury. In a typical case, therefore, the existence of a duty of care will be proved relatively easily.

Argument is more likely to turn on whether the duty was broken; and, if it was, whether the injury was causally linked to such a breach. It will readily be appreciated that both of these matters will turn very much on the evidence available, and the facts that can be proved in Court. In this context, it is perhaps important to stress that the fact that an operation involves some element of risk will not necessarily render an employer liable in negligence. What is necessary is that, where there is a clearly foreseeable risk, then the employer, as a prudent person, should have taken reasonable precautions to prevent others being injured. One guideline frequently used is to ask what is the normal practice in relation to the risk in the relevant industry. Proof that an accident is the result of negligence will usually rest on the *plaintiff* (that is, the person seeking damages), though in some cases the burden of proof moves to the defendant. Since, if negligence is proved, substantial damages may be awarded to the plaintiff, it is now a legal requirement that all employers (save certain exempt employers) be insured against liability for personal injury: the Employers' Liability (Compulsory Insurance) Act 1969.

5.7.2 Breach of the Statutory Duty

Most of the statutory duties outlined above give rise to *criminal sanctions* imposed by the criminal courts—that is, a fine or, more rarely, imprisonment. They are not a basis for a civil court awarding damages for any injuries. However, a separate kind of civil action in tort law may arise where a person is injured, and that person can show that his employer (or other relevant person) had broken a duty arising under a Statute. (Having said this, the general duties arising under the Health and Safety at Work Act 1974, described above (section 5.5.3), will not of themselves give rise to civil liability; this is expressly stated in the legislation. However, given the wide range of more specific duties which appear in Acts of Parliament, and also Regulations made under the authority of the Act of 1974, this is not such a serious limitation as might at first appear.)

The legal rules defining the precise circumstances in which an action for breach of statutory duty can be brought are not wholly free from doubt; there is a lot of law on the subject. For present purposes, however, the project engineer should appreciate that a *breach of duty* may not only render his company and even himself liable to prosecution for a criminal offence, but may also expose him to a civil action for breach of statutory duty to compensate for any injury suffered by an employee (or other relevant victim). Furthermore, it is not permitted for parties to contract out of their liability which may arise if there is a breach of statutory duty. The requirement on employers to insure against liability operates under this head, just as much as it does under the heading of negligence.

5.8 RISK

In this context, *risk* refers to man-made hazards to man, and concerns not just statistics but also attitudes. High levels of risk are acceptable to an individual when the individual is in control of the risk, and not subjected to the risk as say in part of his job. Fischhoff *et al.* (1981) suggest that individuals will subject themselves to voluntary risks that are 1000 times greater than involuntary risks. Another aspect is the number of people that might be affected by an incident—aircraft crashes invariably reach the news headlines because of the numbers involved.

In the past it has often been acceptable for safety measures only to be taken as a result of an incident. This approach obviously becomes unacceptable with large-scale operations, where the cost of an incident (whether human or financial) would be much higher. Nor is it feasible to do everything necessary to prevent every conceivable accident.

Data concerning risks are inevitably statistical, and should always be treated with caution, especially when both the probability and the sample size are low. The following example illustrates how risk can be controlled.

The probability of how likely car brakes are to fail without warning is presented by Brook (1974), along with the possible outcomes; the data are presented here as Table 5.1.

This particular example could be based on a large sample, and the more severe the outcome the more likely there are to be accurate figures

Table 5.1 Probability of Brakes Failing Without Warning

Event	Probability/ Driver/Year	Possible Outcome	Probability of Outcome/ 1000 Cases	Overall Probability/ Driver/Year
Brakes fail without warning	0.005 (that is, once every 200 years per driver)	(a) Car stopped safely on hand brake	900	0.0045
		(b) Minor damage no injury	80	0.0040
		(c) Injury	19	0.000095
		(d) Death	1	0.000005

available. However, it must be very difficult to establish the number of occasions on which car brakes fail with no further consequence.

To reduce the risks associated with brake failure, several strategies could be followed, including:

(1) analyse the failure data and improve the design;
(2) provide monitoring equipment, for example, hydraulic fluid level;
(3) devise a dual circuit braking system.

The final option is likely to be the most expensive, yet it also offers the greatest potential. Maintenance costs will also be increased, and it becomes necessary to check independently each braking circuit. The probability per driver per year of a brake system failure (0.005) will double since there are now two systems, but the probability of this leading to a complete brake failure will be $(0.005)^2$. The probability of the outcome will not be affected, thus the probability of death resulting from a brake failure is $(0.005)^2 \times (0.001)$, not $(0.005 \times 0.001)^2$.

In the UK, legislation requires employers to take all steps that are reasonably practicable to prevent accidents, but some industries are inherently more dangerous than others. Table 5.2 shows the accident rates for different industries, by comparing the incidence of fatalities per 10^8 hours work (this approximates to a 1000 working lives).

On this basis, the incidence of fatalities for air crews appears to be exceptionally high, but it should be remembered that air crews work for fewer hours than other workers. The figures can also be changed by a single major incident. The figures for the chemical industry are based on statistics prior to the Flixborough Accident of 1974 when an explosion of cyclohexane vapour killed 28 people.

The risks associated with employment need to be put into perspective with other risks that individuals take. Kletz (1979) provides the data for ICI shown in Table 5.3.

Table 5.2   Incidence of Fatalities in Different Industries (Kletz, 1979)

Industry	Incidence of Fatalities/ 10^8 Hours Work
Clothing and footwear	0.15
Vehicles	1.3
Chemical industry	4.0
All premises covered by the Factories Act, 1961	4.0
Metal manufacture and shipbuilding	8.0
Agriculture	10.0
Fishing	36.0
Coal mining	14.0
Railway shunting	45.0
Construction erectors	67.0
Air crew	250.0

Table 5.3   Expected Deaths per Year for ICI, 120 000 Workforce (Kletz, 1979)

Cause	No. of Deaths/Year	Comment
Smoking	100	Assume half the workers smoke 20 per day and one-third of such deaths occur before retirement
Influenza	8	Assume one-third of such deaths occur before retirement
Accidents at home	3 or 4	Assume able-bodied people spend eight hours per day at home (excluding time in bed)
Road accidents not at work	15	
Road/air accidents at work or on ICI business	3	
Work accidents caused by chemical hazards	0.75	5.25 accidents in all at work
Other work accidents (falls etc.)	1.5	
Leukaemia	10	Assume leukaemia is equally likely at any age
Heart disease to non-smokers	40	Assume one-third of such deaths occur before retirement, and one-third of the deaths are of non-smokers
All deaths in service as a result of disease	370	

The number of deaths associated with accidents at work accounts for about 1 per cent of the total number of deaths, and the largest constituent is from business travel. The situation is not entirely straightforward, since inter-dependence can exist. For example, exposure to certain chemicals might lead to a greater risk of leukaemia.

The low incidence of fatal accidents is a result of paying attention to safety in order to reduce risks; in fact the risk levels are so low that the cost of saving a further life is very high, especially in comparison with medical examples (table 5.4).

Risk does not solely concern incidents involving people, it also concerns the protection of earnings and assets of a business. These aspects are discussed by Oyez (1978), with particular attention on the petrochemicals industry—an industry that uses large-scale operations with large quantities of materials. The consequences of any incident are obviously greater, but the risks can be reduced by using well-proven methods and new processes that might be inherently safer, such as the use of catalysts to reduce the process temperatures and pressures.

Different aspects of risk include:

(1) injuries to people (employees and third parties);
(2) damage to property (company and third parties);
(3) the interruption of business (and earnings).

A cost can be assigned to all these aspects of risk, and risk reductions could be carried out on an economic basis. However, this assumes a probably unjustifiable degree of confidence in the estimates.

An important stage in risk analysis is the listing of the various hazards in an operation, and here both individual professional knowledge and the use of case studies can be invaluable. Case studies can be found in numerous sources, for example:

(1) technical papers—for example, Wearne (1979), who covers technical and commercial failures;

Table 5.4   The Cost to Save Lives in Medical and Non-medical Situations

Medical	£	Non-medical	£
Lung X-rays for old smokers	1 200	Removal of road hazards	Up to 200 000
Cervical cancer screening	4 200	Removal of hazards in the chemical industry	1 M
Breast cancer screening	12 000		
Provision of intensive care units	15 000	Removal of hazards in the pharmaceutical industry	10 M
Artificial kidneys	28 500	Reinforcement of high rise buildings (after the Ronan Point disaster)	30 M

(2) publications from insurance companies—for example, *Vigilance* published by the National Vulcan Engineering Insurance Group Ltd;
(3) Health and Safety Executive publications.

Since very few failures are caused by hitherto unknown physical phenomena, risks associated with either a plant or process ought to be identifiable. However, familiarity can be an obstruction so there is always scope for an independent technical audit. Finally, accidents do not just happen to other people.

5.9 DESIGN FOR SAFETY

There are two key reasons for considering safety at the design stage. Firstly, many hazards can be avoided or eliminated. Secondly, any guarding or preventive measures can be incorporated at the lowest cost. As an example, a machine imported into the UK required additional guarding that cost about 1 per cent of the total cost. While this is small, it would have been smaller if the guards had been incorporated at the design stage. However, the loss of sales through lost production while a machine is being guarded is much more significant. Safety devices, or guards, are almost inevitably mechanical in nature, since their purpose is to separate an operator from a hazard, be it electrical, mechanical or chemical in nature.

Possible sources of danger include:

(a) electricity (including static electricity);
(b) ionising radiation;
(c) non-ionising radiation (microwaves, ultra-violet light, lasers);
(d) chemicals (whether toxic, explosive, corrosive or flammable);
(e) explosives;
(f) noise and vibration;
(g) pressures and vacuums;
(h) temperatures (either high or low);
(i) dusts (whether from explosions, health hazards or the effects on mechanisms).

The most significant class of industrial accidents is those involving machinery, since this represents about half the total number of industrial accidents. The main emphasis in this section will be on guarding, since the Factory Inspectorate considers that three-quarters of machinery accidents are preventable by sensible (reasonably practical) precautions. Of these machinery accidents about half are caused by employers failing to provide safeguards. The other half are caused by employees removing or abusing safety devices.

Another advantage in considering safety at the design stage is that any guarding designed as part of a system is less likely to interfere with an operator's productivity. It has also been suggested that additional guarding does not reduce productivity, but it may increase the training time and then take longer to achieve maximum productivity.

A question that has to be answered is 'What kind of guarding is necessary'? Lord Cooper gave an important ruling in the Court of Appeal in 1945 that:

> A machine is dangerous if 'in the ordinary course of human affairs danger may reasonably be anticipated from its use unfenced, not only to—
> —the prudent, alert and skilled operative intent upon his task, but also—
> —to the careless and inattentive worker whose inadvertent or indolent conduct may expose him to risk of injury or death from the unguarded part.'

This is the judgement which has provided the criteria for machine guarding in the UK since 1945. Two particularly important points can be drawn from this statement:

(1) 'In the normal course of events' a machine does not run perfectly all the time. Materials or components can break, and operators may be tempted to take dangerous actions that might free the jam.
(2) The final part of the statement recognises the existence of human failings, and states that these need to be considered when the machine is designed. It should not be easy for an operator to remove guards or to defeat safety interlocks.

Before discussing machinery guarding, it will be useful to identify five types of *machine hazard*:

(1) Traps: between parts rotating in opposite directions (gears, rollers etc.); between rotating and tangentially moving parts (belts and pulleys, conveyor belts etc.); between rotating and fixed parts (grinding wheels etc.).
(2) Impact: being struck a blow by moving machinery.
(3) Contact: touching hot, sharp or electrically charged machine parts or materials in the process.
(4) Entanglement: entanglement of hair, gloves, clothing, in moving machinery or materials.
(5) Ejection: flying particles, components or broken machine elements.

Many hazards are associated with normal working conditions, but *contingent dangers* (those that occur when something malfunctions) are equally important. Always ask 'What happens if . . .', this is sometimes called '*Whif*' *analysis*.

Examples of these hazards, and the different strategies for guarding, can be found in BS5304 (1988) Code of Practice for Safety Machinery. This is a very comprehensive document that provides many useful illustrations for practical guarding.

There are many different types of guard, and they will be classified here in order of decreasing safety:

Fixed guards. When properly designed and fitted, a fixed guard will prevent access to all points of danger. The guard must be fixed in such a way that it has to be removed by a tool that is only available to a skilled mechanic.
Interlocked guards. With interlocked guards two important points are: (a) the machine must not be capable of being started until the guard is closed. (b) the guard must not be openable until the machine has come to rest. Indeed it may be necessary to dissipate the potential energy stored in the machine, for example, compressed air in a reservoir. Any interlock needs careful design, whether it is mechanical, electrical or pneumatic. The device should be fail-safe and tamper-proof; this is shown in figure 5.3 for a cam-operated switch. With correct design, any failure of the spring does not enable contact to be made, nor can the circuit be made by pressing the plunger in. This principle needs to be applied in all designs where safety is important.

Magnetic switches, diode links with slugged relays and similar systems all have their advantages and disadvantages, but a very good system is the captive key system. An example of this is where a switch is incorporated into the handle mechanism of a cabinet door.
Automatic guards. These are the type of guard used on some presses and guillotines that push the operator away from the trapping zone as the machine operates.

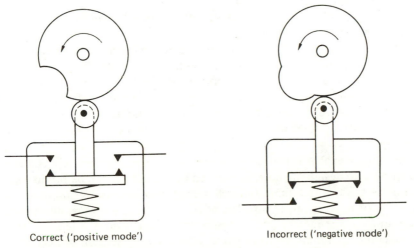

Correct ('positive mode')   Incorrect ('negative mode')

Figure 5.3 Alternative arrangements for limit switches

Trip devices. These are often electromechanical systems that cause the machine to shut down if the operator enters a dangerous zone. Such systems require careful design and installation with regular checking. With this approach, machines need to stop rapidly, and this may necessitate dynamic braking, such as DC injection into the drive motor.

Adjustable guards. This final class of guard is least satisfactory since it gives poor protection and requires resetting; a typical example is the 'fence' on a circular saw. Such guards can jam (or be jammed) in the open position, and even when they are working properly they can enable the operator to contact the hazard.

With the advent of more complex industrial processes and cheaper electronics, digital systems are being used increasingly for process control. The Health and Safety Executive (HSE) uses the term *Programmable Electronic Systems* (PES) to cover all types of computer and controller, namely main-frame computers, minicomputers, microcomputers, Programmable Logic Controllers and Process Controllers. In such systems, Bell (1984) points out that safety aspects must be considered at the outset—otherwise this may lead to extensive (and thus expensive) modifications at a later stage.

Many problems in designing for safety are made worse in a PES. The machine or its program may contain faults or errors, complete testing is difficult, and the many failure modes are difficult to analyse. The small signal levels make the devices susceptible to interference, and faults may occur as a result of transient disturbances. The ease of reprogramming can easily lead to the introduction of errors and, finally, past experience is more restricted. These factors have led the Health and Safety Executive to believe that the assessment of the safety integrity of systems incorporating PES cannot be completely quantified, nor to as high a degree as has traditionally been the case with non-PES-based control systems. The Health and Safety Executive has published guidance documents entitled *Guidance on the safe use of programmable electronic systems.*

In a comprehensive review of machine guarding, Percival (1983) points out that software interlocks are not sufficiently reliable and that a hardware interlock should be used. Robot applications also introduce their own problems, owing to high-speed movements and unpredictable action patterns. Thus most robot applications require a perimeter fence with interlocked access, together with a safe system of work for entry into the enclosure.

Another area that is becoming increasingly significant is noise control. By appropriate design the noise generated by a machine can be minimised, and this is always preferable to such remedial measures as sound-proof enclosures or hearing protection. The effect of noise can also be minimised by attention to the factory layout. Sound-proof enclosures often interfere with machine operation, and supplying workers with hearing protection is not sufficient in itself, an education programme along with a specified system of work are also necessary. A useful description of human hearing and its protection is provided by Lawrence (1978), while Smith (1982) discusses other practical noise control measures.

A key principal in noise reduction is to identify the particular frequencies with an intensity above a datum level, so that the attenuation occurs at the correct frequency. Also, because of the logarithmic response of the ear, it is important to identify the noisiest machine (or part thereof), since by tackling this source the greatest noise reduction can be achieved.

Finally, engineers must remember that it is their responsibility to protect other employees against themselves. It is easy to say when a forgetful or stupid operator injures himself that it is the operator's fault anyway. The engineer may even convince himself that someone else is to blame. However, it is the prevention of injuries that engineers should be concerned about—not avoiding or allocating blame.

5.10 CONCLUSIONS

The change in attitudes and the change in the nature of engineering operations both have an effect on safety requirements. Higher safety standards are expected now, but operations have become larger and more complex, and this makes it more difficult to assure safety. It is no longer acceptable to wait for accidents to occur and then attempt to prevent any repetition.

This theme is also brought out in the Health and Safety at Work Act, in which the object is prevention through participation, rather than just punishment of those responsible for accidents. As with any legislation the position is complex, but Health and Safety legislation is particularly complicated because of the interaction between the many different sources of legislation. This structure has been explained here, along with a description of how the enforcement is obtained.

5.11 DISCUSSION QUESTIONS

1. Describe the historical development of Health and Safety Legislation in the UK. Explain how the Health and Safety at Work Act, 1974 differs from earlier legislation.
2. Discuss the relationships between the Health and Safety Commission, the Health and Safety Executive, and the other organisations responsible for enforcing safety legislation.
3. Distinguish between the meanings of *practicable* and *reasonably practicable*, and the difference between *absolute* duties and those that are *qualified*.

4. On whom are duties imposed by the Health and Safety at Work Act (1974), and what are these duties?

5. What are the powers of investigation and enforcement that are available to the Inspectors of the Health and Safety Executive, and is there any mechanism for appeals?

6. How is a claim for negligence established by an employee? What form does the punishment to an employer take, and can the employee receive any compensation?

7. Identify the factors that affect peoples attitudes towards risks. How can risk be controlled?

8. What are the types of hazard associated with machinery? Identify the different types of guard for controlling these hazards, and identify any shortcomings of the guard systems.

9. Describe the different sources of hazard, and illustrate them with examples.

10. Discuss the reasons why it is difficult to establish the safety of systems that rely on digital electronics.

Part III

Case Studies

Chapter 6
The Planning of
New Facilities

6.1 INTRODUCTION

This chapter treats the planning of a new facility in its widest sense. Most projects will be of a much smaller scale, or be part of a major new operation. However, an overview is always useful, and the principles developed here can also be applied in a reduced form to smaller projects.

Many engineers will be involved in setting up a new facility or operation at some stage in their career. This type of expansion calls for one of three decisions:

(1) whether to expand on the existing site;
(2) whether to seek a new location for the additional facility;
(3) whether to close down the existing facility in favour of a new and better site.

The scope of such a project can thus vary between establishing a plant on a previously unused (*green field*) site, or utilising part of an existing building that has previously been used for another purpose. The options that need to be considered and the decisions that will have to be made obviously vary, but there will be some overlap. To assist with these processes, this chapter includes some checklists. By definition, checklists cannot be exhaustive or cover every possibility; they are included here as a basis for individual adaptation.

When using a green field site, the site selection is of great importance and some of the considerations are discussed in the following section. The geographical location is of major importance since it affects:

(1) the eligibility for national or regional grants;
(2) the quality of transport and other services;
(3) the labour supply and workforce profile.

In comparison, for projects that have to use a site that has previously been used (a *brown field site*), site selection does not arise. Instead, the major consideration is often how to carry out the work with the minimum disruption to other functions. An example of this is the continuous steel casting works at Stocksbridge that are described by Murray (1984). This project involved the part demolition of existing buildings, re-use of some existing equipment, the construction of new buildings and a tunnel underneath a railway line.

Muther (1973) provides a systematic way of analysing whether or not to move an operation to a new site. Muther also adopts a very thorough approach to the planning of new or improved facilities. The stages that Muther advocates are as follows, and it is self-evident that they will also have a profound effect on site selection.

Planning Input Data—what the product is, how it is to be produced, how much is to be produced, what services will be required, and what the project timescale is.

Future requirements—estimates should be made in order to facilitate future expansion.

Interacting components—this refers to items such as layout, materials handling methods and the site services that are needed, in the context of the building.

Component planning—each element of the factory is examined in turn, under the headings of Layout, Handling, Communications, Utilities and Buildings.

Before, throughout and after these processes there will be various stages of planning. These correspond to the conception and evaluation of the project prior to authorisation (see figure 1.1). As time proceeds the detail, accuracy and scope of the planning will increase, until the start of construction. Planning is also needed for the commissioning and the subsequent operation and maintenance of the facility. As with any engineering system, feedback should be used in order to optimise the planning on the current or any subsequent venture. Methods for the planning of facilities are also presented by Wild (1984).

6.2 SITE SELECTION

6.2.1 Introduction

The first decision in site selection is whether to expand on the existing site, or to move wholly or partially to a new site; the new site may be a previously used building, a new building or a clear site. If the existing site can be sold at a good price because it is more valuable to someone else, then this would favour a move. However, if the operation depends on raw materials from the site, or if equipment would be damaged by a move, then a move may not be feasible.

If a decision has been made to move to a new site, there is still the choice of adapting existing buildings that may be available, or building specifically for the new facility. When adapting or rehabilitating an existing building (this is equally applicable to an existing operation) several aspects have to be examined critically:

(1) Is the location really appropriate with regard to markets, suppliers, personnel and transport services?
(2) Is the site large enough for current needs and possible expansion?
(3) Is the building really appropriate? Is the structure sound and intact, is the basic design appropriate, is the age acceptable, is there scope for subsequent expansion or adaptation, is the proposed use compatible with the previous use, are the main services available and adequate, are the amenities acceptable?

When renovating an existing building, it is very easy to under-estimate the cost and time needed for rehabilitation. Also, if renovation is undertaken on an existing facility, then the need for temporary accommodation must not be overlooked, and the cost of lost production must not be under-estimated.

There will also be circumstances that favour a move to a new site; these might include:

(1) if the new site is cheaper, offers better services, has a better environment or has lower overheads;
(2) if the current buildings are unsound, too fragmented or too small with no scope for expansion;
(3) if the existing site services are over-stretched, if a new process requires a radically different layout or if an opportunity is needed to abandon old inefficient practices.

However, care is needed in establishing a facility on a new site, since it is easy to pick the wrong or too small a site, it is easy to under-estimate the time needed for planning and it is easy to over-value the existing site when budgeting for a move.

Many factors will influence the site selection for a new facility, and the order of importance will obviously vary according to different individual needs. Geographical considerations may be important; for example, a process might require a large source of cooling water or hazardous operations might be required to be located away from centres of population.

Transport requirements are also important: rail facilities, main road and motorway connections and airport locations and services all need to be considered. A related issue is the source of raw materials, the destination of the products and the relative disposition of other manufacturing sites and warehouses. When faced with choosing a new operating site, the *Business Location Handbook* (Beacon Publishing) can prove very helpful; the following aspects are covered:

(1) *Transport*. The major roads and motorways are shown on maps, along with details of major road improvement schemes drawn from Department of Transport Publications.

These data are accompanied by an assessment of the effect motorways and major roads have on industrial property. Other sections include air transport (domestic routes and airports handling freight), railway freight services (Speedlink) and waterways; in all cases sources are given for further information. The section on transport also includes comprehensive information on United Kingdom ports, detailing their capacities and capabilities.

(2) *Property*. Regional differences exist for property prices in the commercial, industrial and domestic markets. The *Business Location Handbook* identifies the current differences and provides details of the long-term trends. The variations in rates and rents are also analysed, along with the availability of the different types of business premises.

(3) *Remuneration*. Variations in salary are presented for different regions, but with an emphasis towards clerical and secretarial workers.

(4) *Regional information*. A substantial section is devoted to regional information in the *Business Location Handbook*. The country is divided into regions, and key information is provided about each council and the area it covers. The relevant local government officers are listed, along with profiles of the local workforce, property availability and cost. Details are also given of local communications and a description of the social environment.

There will invariably be other considerations that will influence the site selection; for instance, absenteeism rates vary across the country, and there are also regional variations in the cost of living. Once the location has been established, there may be a choice between adapting an existing building or constructing new buildings. The influencing factors will include the timescale, cost and planning constraints; appropriate advice can be obtained from Chartered Surveyors and Architects.

6.2.2 Minimisation of Transport Costs

Transport cost is evidently not the only criterion for choosing the site of a new facility. However, the minimisation of transport costs may help to eliminate some locations, and provide a short list of possible sites. This approach is likely to be useful when siting a single warehouse relative to its major markets, or when siting a single manufacturing plant in relation to its principal suppliers or customers.

To find the minimum total transport costs, the following expression has to be evaluated for each possible location for the new facility:

$$TC = \sum_{i=1}^{n} T_i Q_i$$

where

TC = Total transport cost to and from the new facility;
T_i = Transport cost per unit quantity moved between an existing site i and the new facility;
Q_i = Quantity to be transported between existing site i and the new facility;
n = Number of existing sites.

Wild (1984) also shows how the optimum location can be found by assuming transport routes to be on a rectangular grid. The advantages of a warehouse sited close to its market arise through savings in bulk transport. If the unit cost between the factory and warehouse is the same as between the warehouse and the customers, then there will be no advantage in a warehouse sited away from the factory. Indeed, the provision of separate site services and management for the warehouse might increase the operating costs.

More often, the problem is not the location of a single facility, but the location of several facilities some of which may already exist. The problem is now concerned with optimising the relative size of the facilities, as well as their locations. This is a complex problem, and the various alternatives have to be evaluated by a linear programming technique. An appropriate technique is the North West Corner Algorithm that is described in Appendix E.

6.3 FINANCE FOR NEW FACILITIES

The discussion here relates to projects in industry outside the public sector and, as outlined in section 1.2, the sources of finance include:

(1) profits from previous years;
(2) equity investment;
(3) debt finance;
(4) Government subsidies and loans.

Where finance is being obtained from outside the organisation, advice should be obtained from an independent source—financial institutions are like any other business and are firstly concerned with their own profitability; if their loan is adequately secured (by a company's assets), they may not be concerned about a project's feasibility or the company's profitability.

The capital in a company can be broadly described as either equity (owners' or shareholders' funds) or as debt. *Equity* is effectively permanent risk capital that neither guarantees a return on the investment not is repayable. In contrast, with *debt finance* there is normally an agreed repayment schedule.

Initially equity is provided by the company founders, and as a company expands and flourishes equity capital is drawn from a wider base. Shares in companies can be sold privately, but companies seeking wider public funding are traded on the Unlisted Securities Market (USM), while large established companies have shares that are listed, and traded at the Stock Exchange. Equity finance is preferable for a firm, since there are no scheduled repayments, or returns on the investment, but in order to attract investment the company has to be financially sound. Investors expect their shares to increase in value, and they also expect the firm to make a profit in order to receive a dividend as shareholders.

Debt finance can take many forms, including loans, overdrafts, bill finance, leasing, hire purchase, factoring and mortgages. However, only *loan finance* is within the scope of this section. Loans are widely described as *short term* (less than 3 years), *medium term* (3–10 years) and *long term* (10–20 years). The source, size and terms for loans obviously vary, but for small firms the loans are almost invariably secured against a firm's assets. If a company fails, the repayment of secured loans takes preference over the unsecured finance (notably the shareholders' funds).

In general, firms will try to maintain a reasonable ratio of debt finance to equity finance, since the latter does not incur interest charges or require repayment. To justify increasing its equity finance (by some form of share issue), a firm should show a good reason (for instance, to finance expansion) to reduce a high debt ratio or to improve efficiency.

The types of finance discussed above are described very comprehensively in, *Money for Business*, published by the Bank of England and the City Communications Centre. This publication also provides comprehensive lists of institutions that specialise in the different types of finance. Brief mention is also made of Government and European loans and grants, but these important sources are discussed more fully in the following books:

(a) *Finance for New Projects in the UK*, published by Peat, Marwick, Mitchell & Co.
(b) *Financial Incentives and Assistance for Industry*, published by Arthur Young McClelland Moores & Co.

The three principal grounds for Government assistance are:
(1) location—for example, development areas;
(2) sector—manufacturing industry, small businesses, new technology;
(3) activity—for example, training, research and development.

In general, Government assistance for industry is managed by the Department of Industry.

Government-assisted areas are shown in figure 6.1; the main forms of assistance for each type of area are shown in table 6.1.

The principal regional assistance schemes are for:

(1) Scotland—Scottish Development Agency, The Highlands and Islands Development Board;
(2) Wales—Welsh Development Agency, Mid Wales Development, Development Corporation for Wales;
(3) Northern Ireland—Industrial Development Board.

In addition, all local authorities in the UK can provide assistance to industry, if it is in the local interest. The assistance can take the form of wage subsidies, rates subsidies, rent subsidies in factories and grants for plant or working capital. Since these schemes are local the terms vary, consequently it is necessary to contact the appropriate Council—an extensive list can be found in the *Business Location Handbook*. The schemes are also advertised in the business section of newspapers.

EEC schemes often support Government schemes, or are implemented through the Government. Specific loan schemes also exist (through the European Coal and Steel Community) for establishing employment in areas where there are redundancies from coal and steel closures. Since loans would be provided in foreign currencies, Government can provide an Exchange Risk Guarantee to cover currency fluctuations.

Business sectors that are eligible for assistance include manufacture; particular examples include projects that might otherwise be located overseas or projects that introduce new products. Schemes also exist for introducing new technology, notably CAD/CAM (Computer Aided Design and Computer Aided Manufacture), FMS (Flexible Manufacturing Systems) and Robots. The nature of the assistance varies and it is often concessionary; the form of the aid can be subsidised feasibility studies, grants towards development costs or capital investment. Certain new industries also attract aid, and these are currently: Micro-electronics, Computers, Telecommunications and Biotechnology. Assistance usually takes the form of contributions to development costs and investment in manufacturing facilities.

The third type of assistance is associated with particular activities: employment, training, redundancy, energy conservation, exporting, marketing, research and development. Details of these schemes can be found in both *Financial Incentives and Assistance for Industry* and *Finance for New Projects in the UK*.

Finally, tax concessions must not be overlooked; businesses must decide how they can most usefully reduce their liability to corporation tax. Special measures exist for private businesses, and individuals can also benefit from

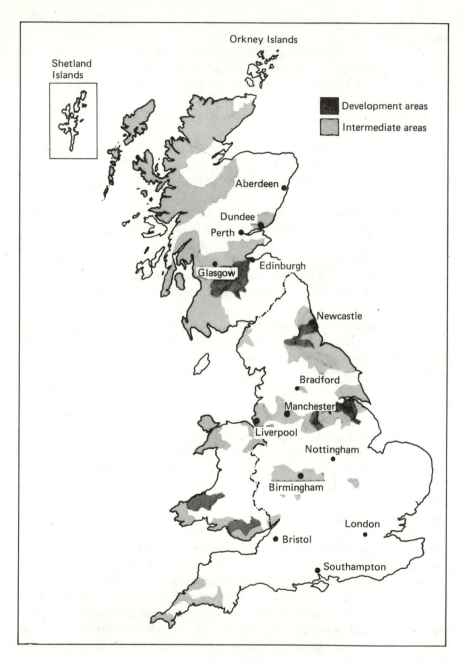

Figure 6.1 Assisted areas in the UK (from November 1984) [Reproduced by courtesy of Arthur Young, Chartered Accountants, London]

Table 6.1 Types of Assisted Area in the UK

Type of Assisted Area	% of Cost of Eligible Assets	Maximum Grant per Job Created (£)
Special Development Area	22	8 000
Development Area	15	5 000
Intermediate Area	0	2 500

tax concessions in the Business Expansion Scheme and employee investment schemes.

6.4 CHECKLISTS FOR NEW FACILITIES

When planning and developing new facilities, many considerations need to be taken into account, and the following checklists raise some of the questions that need to be answered. Such checklists cannot be complete, but they can form the outline for individual elaboration and adaptation.

Site Attributes

Overheads	Rents
	Rates
	Other taxes
Planning Permission	Permitted uses; Town and Country— Use Clauses (Order 1972 lays down 18 different classes of use)
	Neighbouring proposals that may have a conflicting interest; road proposals, outstanding planning applications; scope for future expansion
Geographic	Location with respect to markets
	Location with respect to suppliers
Communications	Roads, parking facilities
	Railways
	Waterways
	Airports
Services	Electricity, substation location
	Gas
	Mains water, boreholes, rivers
	Main drainage, other waste disposal, waste treatment
	Telephone lines
Local Labour Supply	Professional support

	Skilled workers
	Unskilled workers
	Level of unionionisation and demarcation
	Local wage levels
Location Amenities	Housing, supply and cost
	Schools
	Recreation
	Shopping facilities

Building Specifications

Building Requirements	Office space
	Factory space
	Warehouse space
	External facilities
Structural Requirements	Foundations
	Number of floors/floor area
	Floor loading
	Ceiling construction/height
	Roof construction
	Stairways, escalators, lifts
	Windows, fixed/opening doors, internal/external surface finishes, floor coverings
	Roadways
Building Amenities	Heating, ventilation, air conditioning
	Electrical lighting, power
	Washrooms
	Fire prevention and detection
	Resting and canteen facilities
Site Requirements	Landscaping
	Car parking, roads
	Fencing
	Lighting
	Signs
	Security

Manufacturing Considerations

Manufacturing Requirements	Electricity
	Gas, oil
	Water, raw/treated
	Compressed air
	Hot water, steam
	Materials storage
	Materials handling (conveyors,

	overhead cranes)
	Shipping and receiving areas
	Wet areas
	Environment protection (from dust, noise, odours, smoke, vibration)
Production Requirements	Factory layout
	Machine specifications
	Machinery selection
	Machinery compatibility
	Services distribution
	Training
	Maintenance facilities
	Stores
Warehousing	Distribution system—direct to customers or from distribution depots
	Extent of mechanisation for order picking,
	Flexibility in lighting/gangways
Overview	Safety
	Ease of expansion
	Flexibility of layout
	Space utilisation
Personnel	Transfer negotiations
	Additional recruitment
	Job definition
	Salary and wage negotiations
	Training provisions

6.5 CONCLUSIONS

Establishing a new facility is probably the most diverse undertaking in project engineering. Very often a new site will be considered, in which case the transport facilities, local amenities, labour supply and costs, and property costs will all have to be evaluated. If the site has to be chosen to optimise transport costs within an existing network, then linear programming techniques can be applied.

Another major criterion for site selection is the eligibility for grants or subsidies. This is a complex issue, since aid is dependent on location (administered locally and nationally), market sector, the introduction of new technology and activity. Some of these possibilities are discussed in section 6.3, where there are also references to further sources of information. The diversity of financial assistance makes identifying sources a critical aspect of obtaining financial aid.

Finally, some checklists have been presented; since these cannot cover all possibilities, they should be used as a starting point for individuals to construct their own checklists.

6.6   DISCUSSION QUESTIONS

1. What are the circumstances that: (a) favour a company moving to a new site, (b) favour remaining on the current site?
2. Under what circumstances can a firm expect Government aid towards its operations?
3. What geographic and demographic information is needed about a proposed new site for an engineering operation?
4. How can a project for a new facility be broken down, so as to ensure that the project is a success?
5. Under what circumstances do transport costs have a significant bearing on the selection of a new site, and how can these costs be evaluated?
6. What are the types of finance available for a new project, under what circumstances are they available, and what are the respective advantages and disadvantages?

Chapter 7
Computer Projects

7.1 INTRODUCTION

The treatment of a computer system as a project is very little different from the treatment of any other type of project such as building a house, a bridge or a motor car. Only the detailed parts will be different, but like all projects the success depends upon the quality of the project management applied to the project. However, the newness of the technology in computer projects makes the project approach both more important and more difficult; the problems are not of themselves new.

First a definition of terms. The role of the Line Manager was given in section 3.1, and the role of the Project Manager, Project Leader, Project Engineer and Technician were all defined in section 1.1. Additional job titles encountered in computer projects are:

Systems Analyst. This is a term taken from the data processing area, and is the person who investigates a requirement in order to determine the amount of computing required, the size of the computer, the man–hours of design and testing etc.

Systems Designer. This is the person who designs the total software which is generally considered to consist of a number of programs, and their data structures. The system designer is responsible for defining each program, its interface and the data structures.

Programmer. The person who writes and tests the programs to the specification given by the systems designer (this is the software equivalent function of the project engineer).

A distinction has been made between an engineer who designs and constructs hardware, and the programmer who writes and tests the software. Prior to the advent of the microprocessor, most establishments kept to these distinctions. However, since the development of microprocessor systems, many engineers are writing programs and some programmers are putting together hardware systems. There has been much discussion over this matter for several years and some establishments maintain that it is easier to train engineers to become programmers than to train programmers to become engineers; indeed, many of today's programmers started their careers as engineers and have developed the skill of programming. No matter how people are initially trained and educated, both the design of computer hardware and software must be done properly if reliable systems are to be produced.

This chapter is concerned with the engineering and management of projects that require both hardware and software involvement. The amount of each in any one project will obviously vary from project to project and so any statements made must be modified accordingly.

7.2 PROJECT ORGANISATION

Like any project, there must be some organisation as to how the project is to be conducted. The project can be broken down into the following sequence of events:

DEFINITION
SPECIFICATION
DESIGN
ORDERING
CONSTRUCTION/CODING
TESTING
DOCUMENTATION
COMMISSIONING
ACCEPTANCE

This sequence falls into the pattern discussed in Chapter 1, section 1.1, in which the first two steps occur as part of a feasibility study, while the remaining steps, the implementation, occur after the decision to proceed.

The types of projects that are being discussed here are design projects where an artefact of some kind will result. The project may be concerned with designing a new product for a company to produce in production, or it may be a one-off item that has been commissioned by an external company. A systems house or design organisation will undertake projects for customer applications, where the customer will usually be from a different organisation. In this latter case, the systems house will employ sales staff to capture the business.

Apart from internal projects, or products for the company to sell, the project will need to be won in a competitive situation with other companies bidding for the work. In such cases it is typical for the sales staff to look to the potential project staff for help in defining the problem and in estimating the project costs and time. The different types of contract were discussed in Chapter 5, section 5.2.1, and thus a project may be taken on a fixed cost, or a time and materials basis. However, it is not often that contracts are taken on a time and materials basis unless considerable trust exists between the customer and the systems house, and even then, such contracts are usually restricted to feasibility studies. The typical contract is a fixed price contract, and in such cases it is essential to be able to define when a project has been completed. The lack of a proper definition of the completion of a project has caused many a software company to go out of business.

Usually, the project manager is the first person to be employed on a project as he will have a business responsibility for the project, and must therefore be in a position to influence the contract. In any project the first activity is to define exactly what is required by the customer. If the project manager has a good technical awareness and the project is small enough, this activity will be conducted by the project manager in direct contact with the customer. In some cases, the customer is sufficiently technically aware of computer systems to specify exactly what is required. More often than not, however, the customer is not fully technically aware, and may only have a vague awareness of what is required ('I want my factory automated, or I want information about the state of production').

The activity of problem definition is then undertaken, with the situation that the customer knows all about his business and the project manager knows all about computer systems. At the end of the definition stage, the customer and project manager both agree on exactly what the computer system will do. This means that the project manager must get to know a lot about the customer's business, and there will be frequent meetings and discussions to identify what is required.

The definition of the project may be undertaken as part of a paid systems study (time and materials or fixed cost), or it may form the pre-sales activity in which case the cost of the study will usually be added to the tender price. Alternatively, the study could be undertaken once the contract has been won. This latter situation could be very risky if the contract is for a fixed price, as it means that the price was guessed, and the project manager must attempt to meet the customer's requirement and yet still meet the financial targets for his company. The paid study has the advantage of limiting the commitment of both the customer and the supplier, and providing the customer with a written specification of what is required. At the end of the study, the supplier is in a good position to place a realistic price on the job, and has the advantage of being familiar with the system.

Whether the study is done as a pre-sales task, as a paid study or during the project time, the result will be a Functional Specification which will become a contractual commitment on both the customer and the systems house as to what is to be supplied. Both parties should sign the Functional Specification and any deviation from it would require a renegotiation of the contract.

7.3 THE FUNCTIONAL SPECIFICATION

The *Functional Specification* is the most important document in any computer project; it states exactly what the project will do and what it will not do; it will give the limitations to the project and inform the reader exactly what the project is all about. The Functional Specification is written from the users' point of view, and it will treat the project as a black box with interfaces to the user or operator and interfaces to the plant or system that it is controlling. However, it should not give any details as to how the system will be implemented internally. The user is only interested in the behavioural characteristics of the system when connected to his plant, and all operations and descriptions are written in these terms. Many users are naive computer users and the fact that the system has 1 megabyte of RAM with 16K bytes EPROM, memory mapped VDU and a 4 MHz Z80 processor is not only uninformative but is also very confusing, as is any uncommon jargon in the specification. However, the user will be interested in knowing that the system requires $240\,V \pm 10$ per cent at 50 Hz, and will operate at $+10°C$ to $+60°C$ ambient temperature and 90 per cent relative humidity, that the keyboard will have a lifetime of 10^5 key strokes, and that all input voltages must be between $-0.5\,V$ to $+7.0\,V$ etc.

When writing a Functional Specification, the following points should be considered:

(1) The document should inform the user of exactly how the system will behave, how to use the system and its limitations. The intention is that the user should almost never need to ask a question of the system that is not answered by the document.

(2) The system designer should find all the information needed to design the system in the Functional Specification, such that the designer almost never needs to ask a question of the system that is not answered in the document.

(3) The designer should be free to design the system exactly how he wants, within the external constraints of the document.

(4) The document should not contain any facts that cannot be substantiated. For instance, it should not contain phrases like 'the system will be expandable in the future' or 'the system may be modified to provide other facilities'.

(5) The document should be regarded as a legal agreement on what the system will actually do, and may be used by the system designers to

defend themselves (in a court of law if necessary) against any demand made by the customer that was not part of the original agreement. Equally, it may be used by the customer to force the designers to provide any facility that the document says will be provided. For this reason, the document should be vetted legally.

(6) The document must be as watertight and as loophole-free as possible.

Having said this, it is impossible to produce any specification that is 100 per cent watertight and loophole-free that covers every eventuality. However, what is said or not said in the specifications should be binding and any changes should be made on a strictly *quid pro quo* basis with both the customer and the designer happy with the arrangement, and if necessary a change should be made to the contract price.

The Functional Specification can never be completely separated from the design process, as in order to determine what can or cannot be done, or to decide upon certain limitations, some of the basic design must be performed. However, the results of any preliminary design should not be reflected in the specification, and the maximum amount of freedom should be given to the system designers and project engineers.

7.4 PROJECT PLANNING

Once the Functional Specification has been written and agreed, the project plans can be drawn up by the project manager. This will indicate the timing of events within the project, and what resources are required in both man effort and equipment.

It must be remembered that a plan is only as good as the information available at the time of making the plan and, as nobody can predict the future, the plans may well change as more information becomes available. In fact the plans are built upon guesswork, based on the experience of the project manager and others. Nobody can tell what sort of problems will be discovered during design, manufacture or test (there may be a strike or the key designer may fall ill etc.). For this reason the plans and estimating must be done as realistically as possible. It is futile to allow 3 months for a design when everyone predicts it will take 6 months. On the other hand, there may be penalty clauses or contractual agreements to achieve certain points by certain dates.

If the project contains a significant proportion of research work, the possibility must be faced that there may not be an answer to a specific problem or the discovery of the solution may take a very long time. So when a high risk solution is being sought, it is prudent to have a less innovative solution as a fall-back option. All project plans should contain some contingency time as almost every project takes longer than originally predicted. The contingency time must be kept as that and not automatically built into the project time for each function. For instance, if 3 months have been allowed for the design with 1 month's contingency then the design time must not be considered as 4 months—work normally expands to fill the time available. Equally, if 3 months is set for a particular function and that function is completed on time, the end result of that function should be used immediately, otherwise the project manager will lose credibility. The contingency should not be built into the plans but simply be a buffer at the end to accommodate for slippage of the total project.

There are many ways of producing plans (Chapter 2, section 2.2) such as PERT or Critical Path Analysis, but again the results of these plans are only as good as the information available at the time of making them, and plans should always be modified in the light of new information. One very useful planning chart is the bar chart; an example for a computing project is shown in figure 7.1. In any case, the time estimates should be produced by consensus, for the reasons discussed in Chapter 2, section 2.2.2.

From the bar chart, it can instantly be seen when activities must start and finish; for instance, the software testing will probably not be able to commence until the hardware has been built and tested, but some test programs will be needed before the hardware testing can commence. In order to build the hardware by February, some of the components must be ordered before the Christmas holidays; this means the designers must make some guesses as to what components will be required, especially the more esoteric devices. It is sometimes better to over-order at this stage as surplus parts can often be used during manufacture or on another project, and the cost of the components will probably be an insignificant part of the total

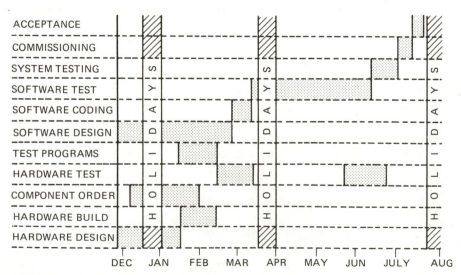

Figure 7.1   A bar chart computer project

project cost. This ordering of inexpensive parts in large batches is an example of the Pareto principle that was mentioned in section 2.3.2.

Once the bar chart has been drawn, a manpower or capacity planning chart can be drawn showing what project staff will be needed and when; this is shown in figure 7.2. Teams for projects involving hardware and software do not generally form at the beginning of the project, and remain together until the end of the project. Instead people come and go for the duration of the project, and only the project manager and project leader will remain throughout.

The manpower chart also shows how many people are required from each line function, and when they are needed. The project manager would then present these to the respective line managers to negotiate staff for the project. This is an example of the matrix management approach which was discussed in Chapter 3, section 3.2.2. The line managers would have similar requests from other projects and would in turn make their own bar charts and manpower charts for their own departments. It may not be possible for instance to have 5 programmers from the middle of February to the middle of March, in which case the project manager must make other arrangements, such as bringing forward certain functions, hiring contract staff, changing the project finish date or negotiating with other project managers for more manpower from their projects.

Account must be taken of the other responsibilities of the project staff, such as commitments to other projects which may overlap the current project. Projects never start and finish cleanly, there is always some pre-sales work, post-commissioning and maintenance work to be done, and a failing system on site nearly always takes precedence over undelivered projects, thus the project manager may not always have full control over the project team. The project manager must also take into account peoples' personal ambitions and development plans. People will be promoted or resign from the company. Some people are better or faster at certain functions than others, but they may not be available for this project. Some people do not

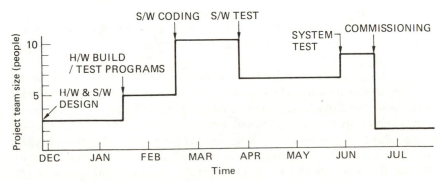

Figure 7.2 A capacity planning chart

work well with certain other people, or may become complacent. All these very human characteristics must be taken into account in the project plans, in the manner described in Chapter 3.

Hardware projects tend to break up naturally into design, build, test and manufacture; but software projects do not break down so neatly and they can extend into considerably longer projects. People working on a function that lasts more than 3 months can easily lose any sense of a target, and it is better in such cases to break the project up into mini-projects, each with its own goal or base level. As each goal is reached, it is as if a new project starts until the next base level. People always work best when they can 'see' the 'milestone' they are aiming at, and have some confidence that they can reach it in the time allowed.

It makes sense for computer projects to use the computer for information gathering and accessing. Electronic mailing and bulletin boards not only make the information available to everyone but also mean that it is automatically logged. A project workbook can be kept on the computer and weekly project meetings should be held for all the project staff, to keep everyone informed and to set goals for the following week. This way information is never more than a week out of date.

The project manager must also be prepared to change the plans on a monthly or quarterly basis, to inform the line managers of any changes in staffing required, and to inform the customer early if there is to be any change in delivery date. (People usually do not get too upset if you tell them 6 months early that the project will be 1 month late, as they can then plan around this, but they do get very upset if they have to phone in on delivery day to find that it will be another month before delivery.)

7.5 DESIGN

By the time a project reaches the design stage, the Functional Specification will have been completed, the contract will have been placed, a project manager selected, a price and delivery date decided and the plans will have been drawn up showing just how much time is available for design. Some of the initial design work will also have been done in order that a project price can be estimated.

The design process should be a *top down* approach, in which an overall structure to the design is decided, the structure is filled with component blocks, and these blocks can then be designed individually. This approach is well known to both programmers and engineers as the best method of designing complex systems. The programmer (or systems designer) decides what modules will be required to implement the function, the interface between the modules and the data structures. Equally, the engineer will take the *black box* approach to the problem and simply sketch out a number of black box modules that fit together as a structure to achieve a basic aim. The programmer and engineer then reduce their modules or black boxes to further, lower-level modules or black boxes within the overall structure, and this process can continue until, at the lowest level, the modules or black boxes become program statements, library functions, integrated circuits or standard circuits.

If a proper structure is used from the start, the whole design process will have the following properties:

(1) a good chance of the design working with only minor modifications;
(2) several people or teams can work on the modules as independent groups;
(3) a readily understandable system, as the basic model of the system will be gained from the top level structure;
(4) a rapid completion.

Having said that the design is performed with a top down type of approach, occasionally it is necessary to adopt a bottom up approach, in order to decide if a particular structure or module really is feasible, or if there is another approach to the problem.

One major problem of design is the 'stuck in a rut' syndrome where an idea has been thought of that will not work but neither can any other approach be thought of. The best solution is usually to stop thinking about the problem or bring in a fresh mind. Modules and structures should be neat and elegant; once they become untidy and complex then there is usually something wrong with the approach, and the design needs rethinking.

The best approach to design is to get other people's opinions. Very often someone will suggest something that you had not thought of before, and this may lead to a better solution. There is no place for pride in design, except in being proud of the final solution, and care must be taken not to reject a solution simply because someone else thought of it—the Not Invented Here (NIH) syndrome. One common technique for problem solving is a *brainstorming* session where a group of people suggest ways of tackling a problem and as a group they use the suggestions to search for a solution.

A true brainstorming session needs rules, and some possible rules are:

(1) have enough people but not too many, 6–12 being a reasonable number;
(2) hold the session in a closed room and not in public, as this will inhibit people;
(3) write down all the suggestions on a board or flip chart for everyone to see;
(4) during the first half, everyone calls out ideas which are all written down and no criticism is allowed;
(5) formulate the question properly rather than vaguely.

At the end of the brainstorming session, the chairman will attempt to group all of the suggestions together to see if any of the grouped ideas can lead to a feasible solution.

It is best if the brainstorming team consists of people from several different disciplines, but without too great a difference in ranks, otherwise subordinates may feel inhibited in putting suggestions forward.

One problem of computer systems is that some decisions need to be made between the hardware/software tradeoffs; this is because most of the functions to be performed can either be implemented in hardware or software. Hardware solutions may be faster to implement and operate, but they will lead to a more costly product in manufacture and be more prone to failure. On the other hand, software often takes longer to design and test than hardware, and so will be more costly to develop. The software solution will also lead to less power being available from the computer. Such decisions are either made by the systems engineer/designer or by negotiation between the hardware and software designers.

Designers not only design a system to meet a functional specification but they also need to produce a design that is manufacturable (possibly by robot or automatic machines), testable (again by automatic testers), service-able and modifiable for the future. The design may be a one-off or it may be a product for quantity production. The problem with one-offs is that very few people other than the designer get involved in the technical details. The system may well work satisfactorily for many years before a problem is encountered, by which time the designer may have moved on to another company or division. The design and documentation must therefore be good enough for others to understand. The problem with production prod-ucts is that it is not too difficult to get one system made and working. However, when several hundred or thousand of the product are being made, all sorts of problems will emerge owing to spreads in component values. The use of worst-case design is very important to ensure that, with all components at their worst tolerance value, the design is still guaranteed to work.

When a design is nearing completion, it is always very easy to take the attitude that it could be done better if it was redesigned. When to stop designing is an engineering decision based on the fact that to redesign something will cost more money, and probably cause the delivery date to slip. This must be weighed up against the benefits of redesigning, such as cheaper to manufacture, more reliable, more elegant, smaller etc. A redesign does not always yield a better solution and sometimes it is better to stick with a known, working design until some major modifications are required.

Some of the design may rely on new products or research ideas, in which case there is a certain amount of risk involved that the product may not be available when needed, or the research will not throw up any answers in time. In this case it is always best to minimise the risk by having alternative solutions available, even though they may be more costly, less elegant etc.

Again an engineering decision may be needed to decide on the possible alternatives, based upon cost, delivery date, ease of manufacture and availability.

Although most companies try and keep up with changes in technology, there are times when new technology must be introduced into the manufacturing areas. An example of this would be the first product to include a microprocessor or logic array. Introducing new technology into a company can cause a number of problems:

(1) educating manufacturing personnel;
(2) tooling costs;
(3) servicing new products;
(4) stocking extra items;
(5) unknown problems.

These tend to cause major repurcussions within any organisation, and careful thought must be given to these problems at an early stage in any project. It may even be the case that an older technology should be used for a small project rather than try and force in a newer technology. On the other hand, if new technology can be introduced on small projects, the teething problems associated with the introduction of that technology can be overcome before a major or more critical project comes along.

7.6 IMPLEMENTATION

Implementation in terms of hardware is the construction of the equipment. There are many techniques of constructing electronic equipment and these may be set by the standards of the company, or even by customer standards. Before construction can commence, hardware must be available, and this usually means parts must be ordered some 6–8 weeks in advance (that is, during design time). The equipment should be made to the highest standards, otherwise problems of confidence will be encountered during testing. The question of 'breadboarding' or prototyping arises with regard to hardware construction. For a device that is to be repeated, in other words be a manufactured device of say, more than 10 units, then it is worthwhile producing a prototype so that problems can be eliminated and modifications performed before the main production commences. During development, if multiple prototypes are needed, the number should be kept to a minimum, otherwise it is difficult to maintain all the prototypes to the same level of modification. For small quantities (1–9) it may not be cost-effective to produce a prototype, in which case care is needed during design to keep the modifications to a minimum. Some modifications are inevitable, and again many companies have standards for implementing modifications. *Breadboarding* is where a circuit is to be tested prior to full construction, to enable component values to be settled, and to form part of the circuit design. Breadboarding is often performed for analogue circuits where there are many unknowns, but there is rarely a need to breadboard digital circuits, as the design process should produce near-correct circuits, and computer simulation can also be applied if necessary.

Software implementation is different from hardware implementation, in that the programs are coded and compiled, and nothing is physically constructed. To perform this task a computer system will be needed, but not necessarily the final project computer system. It may well be the case that the target system will be inadequate for program development purposes. A good *program development system* would consist of a central processing unit with disc storage, visual display unit terminals, printers, a good operating system and support software such as editors, compilers, assemblers etc. The computer does not even need to be the same type of computer as the final target system, as compilers and assemblers are available that run on one computer and produce code for another computer. Testing the programs is however more difficult, and while some simulation work can be done on another computer, the target system will be needed for final system testing. This means that some careful thought must be given to how programs are to be loaded and tested in the target system.

7.7 TESTING

One of the most under-estimated parts of any project is in the area of testing. Testing can be a very difficult task to estimate time for, while some faults or problems can be solved very quickly, others may take a long time.

In terms of hardware, once a device has been constructed there are many unknowns:

(1) the design may not be correct;
(2) the circuit connections may not be the same as in the drawing;
(3) the components may be faulty.

One philosophy of testing is to construct all the hardware and then test the complete system. The other approach is to test each part or module separately, and then test the complete system with some confidence that the individual parts all work. This latter approach is obviously more desirable as the individual modules will be simpler to test, but it may mean that special test equipment will need to be constructed, in which case it may be just as straightforward and cheaper to test the completed unit.

The system will probably be connected to larger equipment or a plant, and it may be totally impracticable or dangerous to test the system on the final equipment or plant, especially in the initial stages of testing. Simulation of the equipment or plant is therefore required, and the level of simulation must be carefully considered early on in the project. The simulation may need equipment to be constructed, and this will have to be done early enough so that it is ready when needed; also the cost of this simulation equipment must be put into the contract price.

Inevitably, the final testing stage will be when the system is connected to the final equipment or plant; this may mean shutting down a factory while testing is performed. This can be a very critical stage in the project as the system is now on site, the customer will be concerned with the time his plant is out of service, and any problems in the system will be visible for the customer to see.

It should be made clear at this stage that there are two requirements in system testing. First there is the functional testing that is needed to prove that the system meets the functional requirements of the project, and the second is to prove that all the parts of the device are working correctly. Strictly speaking, the functional testing need only be performed once, when the system is first built, and from then on only constructional testing is required as the design will not change. However, it is commonplace to conclude the constructional testing by a functional test.

For any system containing a computer, the hardware can be tested by having especially written diagnostic programs; these can be loaded and run on the computer separately from the main system programs. The *diagnostic program* will be written to test and exercise each individual part of the hardware, starting with the local interfaces and working out through all the associated hardware. It may well be the case that some parts of the hardware cannot be directly tested by the computer, only indirectly tested. In such cases, it is common to design in extra (maintenance) hardware that puts the main hardware in a testable state; one typical example would be to form an internal link between inputs and outputs.

The diagnostic software should be capable of testing all of the hardware as a stand-alone system, and it should give meaningful error messages to indicate where and what the fault is. The diagnostic software should also provide 'looping' facilities, so that a particular condition can be set up and left running. This enables a test or service engineer to probe around the circuit with an oscilloscope or other test equipment, to identify the faulty component.

The diagnostic software may well have four parts:

(1) a confidence check to ensure that the overall functions are working;
(2) detailed checks that identify the failing parts;
(3) loop tests to allow engineers to probe the circuits;
(4) a running test that attempts to run the complete system.

In production, the first and last parts would be used on every system, with only the detailed and loop tests being made to faulty systems.

Software testing is a major problem, as software generally has many possible states, and it is usually the case that it is impossible to test the software 100 per cent. A good, structured approach to the design will often eliminate the possibility of many faults, and the use of a good, high-level language with a good compiler can eliminate many more faults. However, there may still be logical faults that cannot be eliminated and so the whole design and test strategy must be carefully thought out. The safety implications of this were discussed in Chapter 5, section 5.9.

The software systems designer will have reduced the software to a number of modules, with defined interfaces and data structures. Each module will have a written specification, and it is part of the process of programming to test the individual modules to ensure that they meet the module specification. This usually means writing special test programs to stimulate the module. Groups of modules can then be tested in similar ways until all the modules are put together and tested, usually with the complete system.

Testing microprocessor systems usually involves both hardware and software testing together. Unlike larger computer systems, microprocessor systems do not have such good development facilities. This means that programs may need developing on other systems that have better facilities, and only then is the final program transferred to the final target system. There are many ways of achieving this program transfer and it is an area that needs careful consideration at the start of the project.

7.8   DOCUMENTATION

Documentation is very important, and yet it is very often given the lowest priority and left until the end of a project. The writing of the documentation should commence at the beginning of a project and continue throughout the life of that project, so that the final system manuals can be produced simply by editing the existing documents. Four types of documentation will be required:

(1) production documentation for internal use only—this should allow the system to be produced directly from the documentation;
(2) the user manuals—these are the customer manuals, and they must contain all the information that the user will require;
(3) service documentation that allows skilled service personnel to maintain the equipment on-site;
(4) marketing documentation—these are the glossy brochures that contain enough information for the technical people, and yet are succinct enough to allow management to decide if it is interested in the product or not.

Most companies have documentation standards and these standards should be adhered to at all times. Standards will be required for:

(1) drawings, including circuit diagrams;
(2) software descriptions and listings (few companies use flowcharts);
(3) manuals specifying the format and layout.

The documentation should be well presented as, like the equipment, it represents the company. Many companies employ technical authors, technical artists, draughtsmen and tracers to maintain the standards of the documentation.

7.9 ACCEPTANCE

Once a system has been designed and constructed to the satisfaction of the system manufacturers, it must be proved to function correctly to the customer, so that the system can be installed and payment made. This acceptance procedure usually takes two forms: an in-house acceptance and an on-site acceptance. The in-house acceptance is usually performed with an acceptable plant simulation, and it is an agreed procedure to be followed (often agreed as part of the Functional Specification, prior to contracts being exchanged). The acceptance procedure will involve both the systems house and the customer, with the project manager ensuring fair play all round. Once a system has been accepted in-house, it can be delivered and installed on-site, where the second level of acceptance is performed. This stage may well be the first time that the system has been connected to the real plant and it can be a very risky and hazardous time. It is essential that the project manager is involved at this stage, as the system may be failing owing to information that the designers did not know about or information not provided by the customer; this could be classed as a breach of contract. Acceptance documents should be signed once a system has been accepted, to enable payment to be obtained.

7.10 SUPPORT

Most systems will be delivered with a period of warranty, and some service and maintenance arrangements. The warranty period may vary from 30 days or less to a year or possibly more, and will cover faulty components or design errors but not necessarily equipment malfunction. Limitation clauses were discussed in Chapter 4, sections 4.6.4 and 4.7. Warranty generally starts from the moment the on-site acceptance documents are signed.

The systems house, having offered a warranty, must be in a position to support the system, and this is again an unknown area. Once a system has been installed on-site, the identification and correction of errors can be a major problem. Indeed it is often a major problem trying to identify whether faults are caused by the installed system or by the customer's own plant or equipment.

Fixing faults on-site can be very expensive for the systems house, as there will not be the back-up facilities available, and there are almost certain to be restrictions on what can and cannot be done with the equipment, both in terms of times of access and which parts can be used. For this reason, most systems houses take extra care to eliminate faults prior to the shipping and installation of the equipment.

As well as the warranty support, the systems house will probably offer on-going service support that is paid for by the customer. There may be many levels of support offered:

(1) full contract cover, where the customer pays an annual fee (usually one-tenth of the system cost) and has service cover with no extra payment—this will usually be offered for 5 day/week 9 am–5 pm or 7 day/week 24 hour cover or some other variation;
(2) call out, where the customer calls in service personnel when required and pays the labour and material costs;
(3) module swap service, where the customer identifies the failing module and returns it with a payment (often one-third of the module cost) and receives a working module in return (often another repaired module);
(4) information only, where the customer supports the equipment and merely requires manuals and spare parts from the systems house.

Maintaining software requires a completely different approach to maintaining hardware, and maintenance usually means fixing design errors as and when they are identified. If software is for general distribution, then a decision must be made whether to update all the software in the field every time an error is identified and remedied. In this respect, the same applies to the hardware if design errors are found.

Most companies will have a formal update or 'change order' system that documents all known faults and fixes to both the hardware and software. Depending on the fault, no action may be taken until a site becomes faulty, in which case the usual policy is to install all the known changes and see if the fault disappears (which it often does). Some failures are either catastrophic, or may cause a safety problem, in which case they would be installed on all sites. In order to keep track of updates, software is usually given version numbers and hardware is given revision numbers. Preventive maintenance may be applied to the hardware where regular servicing is performed, to ensure that failures do not occur.

7.11 CONCLUSIONS

The success of any project depends on the diligence of the project manager. The project manager's role is that of a commitment to deliver a product or service on time, within budget and to the satisfaction of the technical requirements of the project. The project manager also acts as the company's representative to the customer, a single contact to present a coherent and unified policy to the customer, and a buffer between the customer and the design team.

The project manager will preside over progress meetings both internally within the company and with the customer. Monthly customer progress meetings may be held to keep the customer informed of the project progress and to identify possible problems before they occur. If a customer is told early enough of any problems, a solution can usually be found, and there will rarely be any bad feeling.

The project manager is responsible for the cost control of the project, and ensuring that the project is kept within budget, so making a profit for the company. This can only be done if the project manager is kept up to date with the labour and materials costs of the project.

The project manager along with the project leader will be responsible for keeping the project team together and happy, which also means having good inter-personal relationships. The customer, the line managers, the company directors and the project team all interface via the project manager, who must be able to talk to people and resolve personal problems as they occur.

Prior to any project being completed there is often a period known as 'pre-implementation blues', when many people working on the project start to become disheartened. This period can cause problems and one way of avoiding it is to identify milestones on the project, so making mini-projects within the project. Although this does not eliminate *pre-implementation blues*, the level of anxiety is proportional to the size of the project and several small 'blues', spread out in time, are usually better than one 'large' blues that everyone has. It is also worthwhile having a post-mortem at the end of each project to identify good and bad points so that future projects may benefit.

Finally, the project manager must be in the position to decide if and when to pull out of a project, if it is commercially or technically too risky. This decision should not be taken lightly as the repercussions could be tremendous, in terms of customer frustration, penalty clauses and the credibility of the company.

7.12   DISCUSSION QUESTIONS

1. What is the Functional Specification, and how is it produced?
2. How is the top down approach used in design, and what are its advantages in design, manufacture and testing?
3. How does the role of the project manager vary during a computer project? Distinguish clearly between the feasibility and implementation phases.
4. What are the problems encountered in testing electronic and computer based systems, and why are they exacerbated when the system is being used for control purposes?
5. The project leader has just informed the project manager that the original estimates for the project were grossly low. The project has now been running for 3 months and the original estimate of a delivery date of 9 months from the start of the project is wrong. The project leader estimates that a further 12 months are required at the present manning level. Furthermore, the computer is nowhere near powerful enough, and another one is required costing about £50 000. The project cost was £150 000, and the line managers are fully committed for the next 12 months with no obvious slack periods.

   Discuss how such a situation could occur and what possibilities the project manager has available to him. What would be the effect of each possibility discussed?

Chapter 8
A CAD/CAM Case Study

8.1 INTRODUCTION

This chapter is concerned with the selection, justification and implementation of a Computer Aided Design/Computer Aided Manufacture (CAD/CAM) system.

The various processes are illustrated by reference to an example company, and by the implementation that is described below. In addition, the opportunity is taken to discuss alternative strategies for CAD/CAM projects, some of which would have been adopted with hindsight. The project involved initial study, justification and implementation phases, in the manner described in Chapter 1, section 1.1.

8.1.1 The 'Company'

The engineering organisation involved in this project is from the mechanical engineering sector of industry. It is a long established company, now part of a multi-national organisation, with a worldwide reputation within their market. One of the most significant aspects of the company with respect to this project is the importance of the 'Non-standard' and 'Special System' departments. These areas utilise standard products, together with specially designed components to satisfy low volume, often unique, requirements. The annual turnover of the non-standard and special products departments is about £5 million.

The manufacture of the piece parts is by a mixture of conventional metalworking machines, and an increasing proportion of computer numerically controlled (CNC) lathes, milling machines, and a newly acquired sheet metal punching machine. Control tapes for these machines are produced both manually and by using a computer bureau facility.

8.1.2 Initial Study

The questions that have to be answered in this project are: What makes CAD/CAM suitable for a company such as the one described above? Are there universal problems to which CAD/CAM is the solution?

There are some so-called 'classic symptoms' which can help to answer these questions partially, although for a particular justification these may be outweighed by specific criteria. Three of the 'classic symptoms' in this project are discussed below.

Inability of the Drawing Office to React to the Changing Needs of the Manufacturing Areas

In the last three years, the company had worked hard to implement an IBM MAAPICS (MAnufacturing, Accounting, Production, Inventory Control System) system, tied into the sales order entry and invoicing. The results of this were that the production department could react very quickly to engineering changes. However, the drawing office could not, with the effect that components were being manufactured to include changes that the corresponding drawings did not yet show! This was a very poor situation which was a continuing source of problems, especially for the quality control department.

The cause of this was a general reluctance to carry out modifications within a reasonable timescale (there was always something more urgent to do) but more importantly, there was a lack of accountability within the drawing office with regard to the production process. Since the drawing office was traditionally accountable to the technical department, the needs of the production department were never given the attention they warranted.

In addition, a general increase in productivity and efficiency in the factory was still overdue, particularly in the non-standard areas where delivery times were dictated by the overload or backlog of the drawing office.

Transcription of Data

An analysis carried out for a feasibility study on site (table 8.1) showed that significant amounts of time were spent in transcribing existing data from one form to another. For example, a designer/draughtsman would produce an items list (Bill of Materials—BOM) for any new assembly drawing. He would add this to the drawing. Next, a planning engineer would copy out the BOM on to a series of separate sheets (Product Structure

Table 8.1   Potential Main Site Savings

Personnel		Design		Production Engineering	
		9		3	
Total hours anually		15 127		5 175	

Activity	Productivity Gain	Design		Production Engineering	
		Hours	Saving	Hours	Saving
Concept stage	—	759	—	—	—
Layout/scheme creation	1.1:1	2 227	223	—	—
Layout/scheme modifications	1.5:1	759	253	—	—
Detailing	1.5:1	2 277	759	—	—
Searching standard parts	2:1	379	189	—	—
Searching existing drawings	2:1	379	189	—	—
Assembly drawing	2:1	1 138	569	—	—
Detail modifications	1.5:1	2 277	759	—	—
Training	—	379	—	—	—
Checking	1.5:1	379	126	—	—
Schematics	2:1	379	89	—	—
BOM on drawings	1.1:1	759	76	—	—
BOM on file (IBM)	2:1	—	—	1 725	863
NC tape preparation	2.5:1	—	—	2 300	1 380
NC tape proving	1.5:1	—	—	1 150	383
Other productive activities	1.5:1	759	253	—	—
Assistance to others	1.5:1	1 518	379	—	—
(non-productive)	—	759	—	—	—
		15 127	3 864	5 175	2 626
Saving			26 per cent		51 per cent

and Item Master forms), in some cases adding further information. Finally a VDU Operator would enter these data into the MAAPICS system.

Similarly, the CNC programmer would spend a significant proportion of his time redescribing an existing drawing in terms of its co-ordinate data, so that he could then (with or without a computer aid) begin to produce a program to machine a part (a part program). It then had to be typed into a storage facility (in this case a GEISCO beureau) before being output on a punch tape ready for prove-out.

At each stage in both of the two examples above, errors could and would occur. Such errors would eventually be discovered and corrected but, if the entire process (whether BOM or part program) was carried out by re-using the same data (the essence of true CAD/CAM), the errors should be virtually eliminated.

Need to Either Increase Productivity and/or Reduce Manpower

It was felt by the senior management that in order to improve the company's trading position, it should increase the scope of work undertaken by the profitable non-standard area. The continued utilisation in products of a variety of standard parts in many different combinations is of course ideally suited to a CAD system.

Secondly, five of the designer/draughtsmen were approaching retirement age; the introduction of a CAD system at this point should allow sufficient time to draw on their wealth of product knowledge, and make their replacement on a man-for-man basis unnecessary.

The 'D' in CAD is often mistaken for Draughting rather than Design, probably because there are suprisingly few CAD/CAM installations where there is a formal design need (for example, stress analysis, dynamics etc.). This is certainly the case in the example company, owing perhaps to the intuitive nature of the engineering, and the experience of its design and development engineers.

Having recognised the existence of some of these classic symptoms, an initial feasibility study was undertaken, to establish the likely basis for a justification. A team of evaluators was founded, drawing upon the expertise of the Group CAD/CAM Manager and containing representatives from each engineering department. This is an example of the secondment model described in Chapter 3, section 3.2.2.

8.2 THE JUSTIFICATION

A justification of this type has four distinct areas, the first two of which are mandatory in order to gain approval while the final two are not:

(a) Technical;
(b) Financial;
(c) Management;
(d) Users.

The two latter categories are quite intangible; however, with hindsight for a project of this nature, they are just as essential as the mandatory sections of a justification. Indeed, ultimately an installation will stand or fall by the ability and attitude of its operators, and this should never be forgotten.

The approach to each justification and its particular problems will now be discussed in turn.

8.2.1 Technical Justification

The technical justification was divided into two parts: vendor selection, and benchmarking.

Vendor Selection

Various CAD/CAM handbooks are now available to enable prospective purchasers to draw up a 'long list' of possible vendors. Such reference guides are generally segregated by application (mechanical, microelectronic etc.), system size (single user, networkable etc.) and, of course, cost. In addition to these, engineering exhibitions provide a showcase for the major vendors.

Another popular approach is to enlist the assistance of consultants in both the pre-installation and post-installation phases. This approach was resisted in the example case because it was felt to restrict the development of in-house expertise.

The 'long list' can be reduced to say four or five serious contenders by a combination of standard demonstrations (at exhibitions or vendors' premises) and perhaps visits to existing users of CAD/CAM. The resulting 'short list' of vendors can then be examined in much greater detail to establish the preferred choice.

Elements of the detailed evaluation would include:

(1) vendors' commercial and financial stability;
(2) on-going development plans;
(3) training and support organisation;
(4) number of installations in similar businesses;
(5) initial configuration and future expansion;
(6) comparative costs.

Visits to current users need to be viewed with some reservation, since after all, the vendor would not take you there if anything other than a favourable reception was to be had. However, such visits are useful, especially for assessing the resources and facilities needed to manage such a system, as well as to pick up useful tips.

A visit was made to one of Applicon's Marconi sites at Rochester in Kent. Mechanical and electrical software was demonstrated at an installation with three separate systems. The visit confirmed Applicon's suitability in the design environment, but it was not possible to see any manufacturing software in use. Under normal circumstances, the lack of existing users would and should cause alarm (how competent is the software?). A look at the Applicon userbase showed its mechanical engineering customers to be predominantly of the Marconi type (that is, design and assembly) but with sub-contracted machining. Since Applicon is part of the Schlumberger

Group, it satisfied all the criteria necessary to ensure a continued presence in the CAD/CAM environment.

Table 8.2 illustrates the five selected vendors for the example case. The system offered by Applicon was generally favoured by the evaluators, and so it was decided to recommend the purchase of that system, provided that it satisfied a formal benchmark.

Benchmarking

Benchmarking is an accepted method of 'test driving' a CAD/CAM system. It can also be used to provide an abundance of other useful information about a potential vendor, beyond that designed by the benchmark itself. The example benchmark consisted of two exercises: one a design, the other the production of a machine control tape for a CNC sheet metal punching machine. However, benchmarking can be quite an emotive subject. From the vendor's point of view, it is very time-consuming and labour-intensive, particularly if it is of several days duration. ('What do you want to benchmark for? We wouldn't sell the system if it didn't do what we said it would'.)

When defining a benchmark, be realistic; that is, address only areas which are of primary interest to your business. Base this on an analysis of the likely requirements in the next 2–5 years. Concentrate on highlighting any aspects which are critical to your requirements. For example, with numerical control (NC) is it really necessary to have a 5 axes machining

Table 8.2 Budgetary CAD/CAM Configurations

	Cost (£) CPU with 4 Workstations At Main Site, 1 Each At Two Remote Sites				
	Kongsberg	Calma	Gerber	Applicon	Counting House
Host workstation	150 000	180 000	200 000	194 000	170 000
2 Remote workstations	135 000	90 000	120 000	128 000	143 000
1 Plotter	20 000	20 000	20 000	20 000	20 000
Software	35 000	50 000	20 000	33 000	—
Delivery, installation, import duty etc.	15 000	30 000	—	—	30 000
Maintenance	20 000	30 000	35 000	—	30 000
Training	10 000	10 000	—	—	10 000
Sub-total	385 000	410 000	395 000	375 000	403 000
DOI grant	(60 000)	(60 000)	(60 000)	(60 000)	(60 000)
	325 000	350 000	335 000	315 000	343 000

capability, or will 3 axes be sufficient? There is generally a significant cost penalty for increasing the requirement beyond that actually necessary.

On top of this, there is a potentially serious risk to an over-complex benchmark. It may well identify a particular supplier as clearly superior at this type of work. However, that system might well be quite ordinary or even inferior in dealing with the day-to-day work of the company. A realistic benchmark will at least ensure that the chosen system is the best one for the work it is most likely to do.

Other useful information can be gained from the way in which the benchmark is performed. Watch for excessive use of 'canned cycles' (typically activated by a function button). These generally indicate that the required action is not a standard feature of the system—how much time and effort went into creating this facility? The ability to customise and refine the system is essential to produce maximum benefit, but do you have the resources or time to customise it at such a basic level, especially if it also requires special skills such as the ability to work with Fortran?

Observe the user interface closely—that is, how easy is the system to drive? Excessive use of the keyboard is to be avoided unless the prospective operators are good typists. During the benchmark, pay particular attention to the competence, speed and skill of the vendor's operators. This should be indicative of the level of competence that a full-time operator should achieve; if the vendor's operators cannot reach acceptable operating levels (particularly during the unplanned parts of the benchmark) what chance do your operators stand?

In addition, look for and ask about the minimum level of expertise needed to drive the system. The ability of a 'casual user' to operate the system may be important, especially if the system is to be used by engineers (for analysis) who certainly will not be dedicated to the system on a full-time basis.

Bearing this in mind, very close attention was paid to the punching part of the benchmark, since this was to be a significant part of the financial justification. Applicon agreed to support the punching post-processor from their Stockport Office (all other software problems are fixed in the USA) and Applicon showed sufficient commitment to the CAM field to negate the lack of any existing UK users. The outcome of the technical investigation was that Applicon satisfied the foreseen requirements in the CAD/CAM field.

8.2.2 Financial Justification

To take advantage of a special offer from Applicon, it was decided to purchase a system with four graphics workstations—this being the full complement supported by the main processor. The cost of this system was now £198 000, as opposed to the figure of £375 000 shown in table 8.2.

The capital cost of the CAD/CAM system was offset against both an internal saving and an external grant. The internal saving was available from the CNC punching machine installation. Here part of the project cost was for a control tape preparation facility. This would not need to be spent if the CAD/CAM system was acquired. The external grant was a Department of Industry (DOI) grant from its CAD/CAM awareness fund (as mentioned in Chapter 4, section 4.3). The fixed asset expenditure and the estimated annual revenue costs were as follows:

Fixed Asset Expenditure

	Savings £(000)	Capital Expenditure £(000)
Project Capital Cost		198
Already Justified for		
CNC Punching Machine	27	
DOI Grant	60	
	87	87
Net Cost		111

Annual Revenue Cost

	Savings £(000)	Costs £(000)
Hardware and Software		
Maintenance Cost		14.6
Programming Bureau Saving	7	
Illustration Saving	8	
	15	15
		(0.4)

The net saving in revenue was seen to be negligible.

The exact nature of producing a justification will be very dependent upon the particular business organisation concerned. Each company will have its own accounting methods, and these will greatly influence the way in which the justification is produced. Indeed, the expenditures given above have already been conditioned by including initial software costs as capital investment, whereas it is quite common to find software costs absorbed as revenue.

Despite the unique nature of a justification, there are several basic approaches which can be adopted, and it is useful to review them here.

The financial case is relatively straightforward if the business requirement is for complex prototypes or 'one-offs' where design or production errors cause scrap or waste of several thousand pounds at a time. Similarly, if printed circuit boards limit the design process, then automation can cut the sub-contract and design times by as much as 80 per cent. In such cases, a typical justification might take the following form:

	£(000)
Typical prototype costs per project	15
Typical scrap or waste per project	5
	20
For 10 projects/year (at £20 000 each)	£200 000

against £500 000 capital investment
payback period three years

The justification is not so straightforward if, like the example company, this mode of analysis is not applicable.

A more traditional approach, and an option explored in the case study, is to attempt to equate the productivity gain to a reduction in manpower. Consider the savings for the site as set down in table 8.1.

	Hours saved per year	Men saved
Design	3864	2.35
Production	2626	1.51
		4

For an employment cost of £10 000 per annum, a yearly saving of £40 000 will result. However, this would need to be offset against a one-off redundancy cost of say £40 000. With a discounted cash flow analysis (see Appendix A), the savings give a mediocre internal rate of return of less than 15 per cent.

In addition, this approach is quite artificial on a practical basis. For example, in the production engineering area, the savings would allow the release of one NC programmer. There are only two NC programmers in the department; of course with holidays, sickness etc. it is not possible to operate with a single programmer. In effect this saving is nullified unless there is a high level of job flexibility.

The above examples of traditional justification methods, as applied for example to machine tool purchase, are not strictly applicable to a justification for a computer based system. It is of course possible that they might produce favourable returns but, given the somewhat unique nature of engineering computer capital justification, there are alternative approaches which may prove more fruitful.

In reality, a computer based system will become technologically outdated and inefficient within 3–5 years at the most. A justification ought to reflect this short life, and make payback calculations accordingly. To the parent of the example company, the method of justification remained the same for computer equipment as for traditional capital equipment, because probably the financial management personnel were still largely unfamiliar with dealing with such large-scale computer purchases. So the company were forced into adopting a traditional approach to the justification, since this offered the only real chance of approval within the required timescales.

In fact, if the justification had been based on a five year payback, rather than ten years, then it is doubtful if the purchase would have been approved. The main reasons for the poor financial justification of the system on this project were:

(1) The company did not produce expensive (or cheap) prototypes on anything like a regular basis, so there were no savings to be made from 'software prototyping'.

(2) Very little formal design analysis took place on the company products, so there was no benefit to be gained immediately from say rotational, stress or thermal analysis. Undoubtedly benefits could be gained, but these could not be a primary source of fiscal justification.

(3) There was some immediate benefit to be gained from having CAD/CAM in-house, by closing the existing NC bureau facility and holding the library on it; the saving would be around £7 000 per annum.

(4) The only other immediate source of fiscal saving was by a reduction in staff, directly attributable to an increase in productivity; something accountants of all levels can relate to, as opposed to the myriad of intangible benefits from CAD/CAM like integrated databases, re-use of data, reduction in errors etc.

Against this background, to achieve an acceptable internal rate of return on the project (that is, 17 per cent) a saving of five people was required. The figures in table 8.1 show that using conservative but realistic CAD productivity data, it would only be possible to predict a saving of four people (26 per cent of 9 and 51 per cent of 3). As explained in section 8.2.2, this approach has many shortcomings and is really impractical.

Hence to achieve a saving of five, a different approach was necessary. The concept of 'CAD Engineers' capable of multi-disciplinary activities was adopted, to allow the necessary savings to be made.

These engineers would be capable of activities spanning several traditional roles. The possibility only becomes practical because of the de-specialising nature of CAD/CAM. The dramatic impact on the Non-Standard Products Department can be visualised with the aid of figure 8.1a and b.

As already mentioned, this was an area in which maximum benefit could be gained from the introduction of CAD/CAM. However, as can be seen from figure 8.1a the existing function of the department did not easily lend itself to maximising these gains. An incoming enquiry for equipment would be passed to a contracts engineer. The contracts engineer, acting as a co-ordinator, would begin the process of producing a formal quotation. By calling on the help of both the technical department and an estimating engineer, a suitable technical configuration would be chosen. A desig-

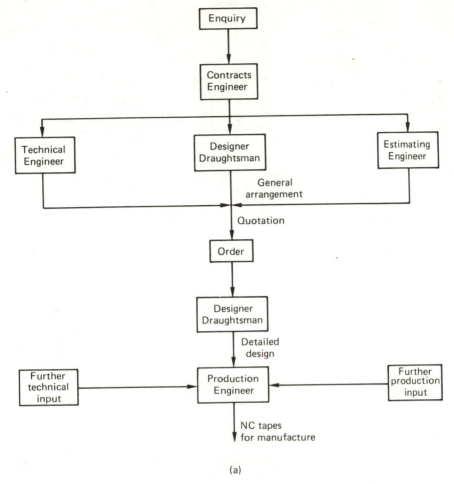

(a)

Figure 8.1 (a) Traditional operation of the Non-standard Products Department

ner/draghtsman would then produce a general arrangement drawing of this configuration, and a formal quotation would be sent to the customer.

Upon receipt of an order, the contracts engineer would then assume project responsibility. With further technical and production aid, detailed design would begin. When detailed design was completed, the drawings were handed to production engineers, who decided which items were suitable for NC manufacture and produced corresponding control tapes.

The characteristics of such an operation are the number of specialists needed (estimator, designer, technical, contracts, production). Each of these could perhaps benefit from a CAD/CAM system to some extent, but to realise its true potential an organisation such as that of figure 8.1b is more suitable.

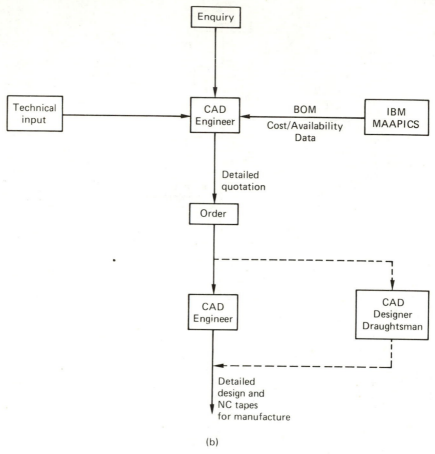

(b)

Figure 8.1 (b) Operation of the Non-standard Products Department utilising CAD/CAM

Here the central figure is a CAD Engineer. He receives an enquiry, and using existing information on the CAD system he puts together a design, calling upon the technical department if necessary. Then using a link to the production control system, a bill of materials is quickly transformed into standard cost data. This leaves only the truly non-standard elements of the design to be costed before a bottom line total cost is calculated. Furthermore, the quality and presentation of the quotation to the customer is greatly enhanced.

Upon receipt of the order, the CAD engineer can complete the design, passing some of the more trivial sections to the designer/draughtsmen during busy periods. The complete design would now include NC tape preparation on the CAD/CAM system.

Of course, the CAD engineer in this role must possess some special qualities, namely a knowledge of contracts, estimating and production techniques. The facilities provided by a well engineered and organised CAD/CAM system would facilitate this (with the exception of contracts), removing the need for pure specialists and replacing them with well-qualified, experienced engineers instead. The benefits from each of the areas within the non-standard products department could be accumulated by amalgamating the roles of some of the engineers, with the CAD/CAM system handling the mundane tasks.

Senior management were convinced that such an organisation would function very successfully on the site. Indeed, certain engineers were identified as prospective 'CAD Engineers'. With such an organisation there was very little hesitation in proposing a cut in staff of five following the introduction of the CAD/CAM system. At the time, it was not possible to forecast where these five losses would be made, but the non-standard products area was likely to account for at least three of these.

For an organisation like the one just described to function effectively, it would require the CAD/CAM system to be on stream with a significant database of parts already available. This would represent a situation typically 18 months from installation, so the commitment to lose staff was made for a period about 18 months to two years from the installation of the system.

Given the two year time delay between installation and release of staff, it is possible that some, if not all, of the five individuals would leave the company anyway. Thus it would not be necessary to replace them, nor would it be necessary to incur the redundancy costs of such an action. Financial analysis was thus carried out for two cases: (a) with the full cost of redundancy and (b) with no redundancy costs.

One final approach to the financial justification would have placed the direct cost of the project on the Non-Standard Products Department. Since the department was identified as a prime source of the CAD workload, some fiscal return might have been realised in the following way:

Non-standard business currently handled by 5 designer/draughtsmen	value £5M approx
Potential enquiries/tenders not quoted because of over-load or timescale restrictions (that is, additional business lost)	value £500k approx.
Presence of CAD/CAM would (conservatively) allow the same personnel to quote for 60 per cent of the lost business, and if only 60 per cent of this was realised (that is, extra business)	value £180k approx.
the high profit margin on the non-standard business would increase profits significantly	increased profits £100k

Such an approach would also have carried the advantage of promoting extra business, rather than the reduction in personnel, which always seems to be associated with new technology and creates resentment. Interestingly, the figure of £100k represents nearly the net project spend on CAD/CAM for the installation, and although it would not be realisable in the first year, it is believed to offer a realistic alternative justification.

8.2.3   Managerial Justification

In the specific example here, the CAD/CAM justification was somewhat unorthodox, so it was essential to secure the backing of senior management. Indeed, it is necessary to ensure management approval for all projects, but especially those where the financial case is not strong. Positive management backing can be the difference between success and failure to obtain approval.

Such persuasion is undoubtedly a form of propaganda, but its success is a big step towards approval of the justification. Production and technical management (whether or not already involved) should be easily won over for obvious reasons. However, it is important that 'management selling' does not stop there—just about every senior manager can be made to see great benefit for himself and the company from the introduction of CAD/CAM.

For example, the Sales Director should see the benefits to the company image, as well as in improving the quality and speed of the production of quotations to his customers. The Personnel Director should see the ability of new technology to attract engineers as a benefit.

One particular manager whose support is vital is the Data Processing Manager (often called the Management Information Manager). As a result of his computer expertise (even if unrelated) he will be consulted on technical matters, such as interfacing with the commercial computer system. Any negative reaction on this front will result in at best a change in vendor, or at worst a rejection of the project.

If it can be recognised that any member of the senior management is particularly influential in setting company policy, then it should be possible to take advantage of this. By using behind the scenes tactics, and a justification weighted towards the benefits to that individual the chances of successful approval are increased considerably!

Another neglected part of the management justification concerns setting realistic goals for the project. A successful CAD/CAM system will bring significant benefits, but it is important for management not to expect returns in too short a timescale. Experience has shown that it may take typically 12–18 months for some of the benefits to become realisable. In the meantime there will be problems and disruption caused by the implementation. The significance of the disruption and disillusionment can be minimised by pre-empting the situation with practical, realistic goals, as described in Chapter 7, section 7.11.

Of course, the timing of this conditioning is crucial—it is undesirable to dampen enthusiasm, but it is just as undesirable to gain approval for a project doomed to failure by unachievable objectives.

8.2.4   User Justification

Users or operators of a CAD/CAM system are such an intangible group that in virtually every justification they are ignored. However, at the end of the day, the ultimate success or failure of such a project depends upon the people who use the system. So, in a similar vein to the managerial justification, it is important that the users appreciate the value of the system. Moreover, because in the vast majority of cases the threat of redundancy hangs over a CAD/CAM installation, the potential users are naturally apprehensive.

This apprehension is accompanied by rejection from those who consider themselves vulnerable to new technology. In the example case study, the users were neglected. Reference was only made to them following project approval. They were informed as a matter of course during the negotiations to place the contract. Again, experience showed that many problems, especially those relating to motivation and attitude, were attributable to the feeling of a lack of consultation by the users.

In any prospective installation this situation can and must be prevented. The most logical way is by including the likely users in the evaluation team, even if only in the preliminary stages. The resulting feeling of participation can only be beneficial, and a positive attitude should result.

8.3   PLANNING FOR INSTALLATION

8.3.1   Formation of a Project Team

As with system selection, the implementation phase is best managed by the use of a project team. This team may or may not include members of the evaluation team, but it should include members from every area to be directly involved with the system. The team should be headed by a senior manager and should contain at least two engineers charged with developing the installation and providing in-house expertise.

To this end, it is essential that the key members of the team receive pre-installation training. This will provide an insight into the use, setup and control of the CAD/CAM system, without which it is impossible to pre-plan the installation.

8.3.2   Training Programme

One of the first tasks to be undertaken is to establish a formal training schedule with the vendor concerned. System purchase will have included a variety of training modules in all aspects of the system: management, basic user, applications etc.

A typical bar chart (Chapter 2, section 2.2.2) training schedule is shown in figure 8.2.

The key aspects of the programme include:

(1) Adequate pre-installation (and therefore offsite) training for the implementors, so that preliminary work such as account allocation can take place.
(2) Basic user training just before installation, so that from the first day following commissioning the system can be used.
(3) Careful selection of probable users.
(4) Adjustment of users' workloads to provide sufficient 'hands on' experience outside the formal training courses. Serious consideration should be given to the use of contract labout to reduce the day-to-day pressure on users.
(5) A realistic timescale—do not be over ambitious when setting the dates for courses. Resist the temptation to condense all the courses into the shortest timescale possible. It is quite important to allow periods for consolidation and to avoid the feeling of 'perpetual training'.
(6) Use of off-site training locations whenever possible (such as at the vendor's training school). The atmosphere and special-purpose facilities

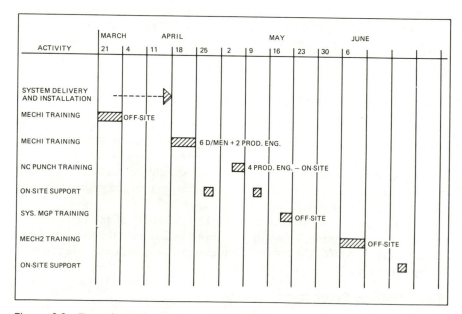

Figure 8.2 Typical implementation schedule

are more conducive to learning, and there is less likelihood of interruption from telephone calls etc. These advantages outweigh the additional costs (for example, accommodation) which are associated with this approach.

(7) At every opportunity during training, introduce company work, rather than relying on vendor-supplied material. This provides benefits in two areas. Firstly, there is more incentive for the operators to complete work if it is not just meaningless drawing; and secondly, it is productive work—that is, the finished drawing can be used and it represents an early success for the CAD/CAM system.

(8) Establish a contingency plan. This should provide for as many changes in circumstances as possible. For example, if it becomes necessary, can the training schedule be easily delayed? What happens if the selected users fail to come to terms with the system? Ideally the plan should never be needed, but the quality and comprehensive nature of such a plan may be critical to the success of the whole project. The importance of contingency plans was mentioned in Chapter 3, section 3.2.4; their presence helps to rebuild morale after any setbacks.

8.3.3 Siting the Hardware

There are several important issues to consider at this stage. It is probable that a special-purpose computer room already exists for the data processing department, and this is often the location that is considered first for installation of the main processor, disc drive and even the plotter.

The attractions of this solution include the availability of the ideal environment, as well as personnel to handle day-to-day operations such as maintaining the backup and archive copies. However, against this must be weighed the disadvantages, which probably include restricted access and a remote location (both a problem for plotting) and also a lack of knowledge concerning technical computing.

A more popular approach is for the system to be run (albeit ultimately) by the operators, and so the system can be sited close to the engineering departments. This was certainly the case in the example company.

The location of the actual graphics terminals raises separate issues. If the terminals have been justified to carry out specific tasks (such as NC tape production), then why not site them in the appropriate department? There was considerable pressure to allocate terminals on this basis in the case study, in which case the allocation would have been as follows:

Systems Design	1 Terminal
Component Design	1 Terminal
Production Engineering	1 Terminal
Technical Department	1 Terminal

In essence this allocation provides for all prospective users of the system, but it is in reality quite weak and inflexible. For example, although the systems and components design could utilise a terminal each on a permanent basis, the technical department could not, and although the initial load from production engineering might even exceed a single terminal, it would eventually diminish to represent less than one.

Under such circumstances, maximum utilisation of the terminals would be unlikely, and because of the remoteness of the terminals, users from one department would be reluctant to use terminals in another. It would also inhibit self-help and cross-fertilisation of ideas.

8.3.4 Personnel Selection

It is rarely possible to train all the likely candidates in one course, not least because it is impractical to remove all the designer/draughtsmen from their work for a whole week.

Certain engineers may well be natural and automatic choices, as with the engineer responsible for the NC punching machine in the case study. But selection of the remaining trainees may have to be made on a more arbitrary basis; the guidelines developed in Chapter 3, section 3.4.2, may also be useful.

As a general case, experience has shown that young people do tend to adapt to CAD/CAM systems in particular, and computers in general, faster than the older generation. However, this is a very sweeping statement and it should not be used as the sole criterion. Instead, a combination of factors should be used to select trainees:

(1) If the system is to be manned by dedicated (rather than or in addition to casual) users, then these individuals must receive formal training. They should also be assigned to the system as early as possible.
(2) Resist the temptation to give supervisory personnel (such as drawing office supervisors) places on the user training courses at the expense of 'real' users. This is wasteful of training resources. Instead, if the vendor does not already offer such a course, press him to offer a 'system appreciation' course. This should include some 'hands on' experience, but it should also aim to increase awareness rather than user skills. It is a good idea to include invitations to Trade Union/shop floor personnel to this or a similar internally run course.
(3) Look for a positive (even enthusiastic) attitude from candidates. Individuals with these traits will persevere and succeed. It is better to invest training resources here than in those who express negative or resentful attitudes.

The above criteria are by no means exhaustive, but they were all overlooked at the example company. Personnel selection there was made on the basis of seniority. Supervisory staff were trained at the expense of identified dedicated users. The attitude of the majority of these people was also less than ideal; thus it was not surprising that the training phase was very disappointing. The result was that on the completion of formal training, there were insufficient competent operators to utilise the system fully.

8.4 COMMISSIONING

The duration of the commissioning phase of a CAD/CAM system will obviously depend upon the size of the installation. Even though the duration may vary enormously, the same basic characteristics can be identified in every installation.

8.4.1 Building the Database of Parts

In order to realise the maximum benefit from the design side of the system, so-called 'standard parts' such as nuts, bolts, connectors etc. must be available 'at the touch of a button'. As well as these, standard manufactured items, both internally produced and bought out, should also be available.

For optimum throughput, a database of several thousand items is not uncommon. In itself this poses a number of questions:

(1) Where to start?
(2) How should the data be organised?
(3) How should the parts be represented graphically?
(4) Is there sufficient disc storage to hold everything on-line?

Again the answers depend on the application, but broad guidelines can be drawn.

Where to start?

There are two fundamental choices to be made on starting the database. One option would be to attempt rapidly (but systematically) to add all the parts to the database—for example, nuts, bolts, washers, then all the connectors, flanges etc. On the surface this perhaps looks attractive but it has a number of shortcomings.

Firstly, operators dedicated to building libraries of parts are not directly productive. At the end of several days or weeks of work there will be no obvious output. It is also possible that many of the items will not be required for a lengthy period of time. Secondly, both the quality and quantity of operator output will diminish markedly when given this type of task. This is almost certainly due to the repetitive, non-motivating nature of the work.

An alternative and more constructive approach is to phase the buildup of parts. The ideal commissioning phase (as discussed later) identifies a workload of 'live' tasks for the operators—for example, new design projects. To reach completion these will require the creation of several standard parts. As they are created, these items are added to a suitable library for more widespread use. This method is productive in two ways; useful, necessary work is produced and no effort is wasted by creating parts not immediately required.

Given the nature of the company in the case study (that is, a non-standard business utilising a high percentage of company manufactured products) the second option was the obvious choice. For each specific project, the standard parts were identified and if not already available, they were constructed during the project and added to the library upon completion.

How should the data be organised?

This topic is especially difficult to generalise upon, because it is not only system user dependent, but also system vendor dependent.

It is usual for a CAD/CAM system to support more than one library, and if this is the case, then it is normal to have separate libraries with similar items in the same library. This is particularly easy if parts are coded using say BRISCH or parity codes, whereby similar parts have nearly identical numbers; such systems are described by Wild (1984).

Again most systems provide for a hierarchical data structure such as that shown in figure 8.3. This structure of three tiers is organised on a project basis, and as such is dynamic. New branches would be added as a project starts and then deleted upon its completion and subsequently stored in the archive.

From the figure it can be seen that the three tiers provide three levels of library available to the users. The availability of parts from a particular library increases near to the top of the tree, as does the limitation on access for change. In this way each user can have a library local to a particular project. He is completely responsible for its upkeep, and has the ability to change parts in this library; this would typically be a development library.

At the second tier would be held project specific libraries of parts. These would be available for use by all members of the project but could only be changed by the project supervisor.

Finally, at the top level would be company-wide libraries. Any user would have the ability to use a part held at this level. Modification to parts at this level could have significant knock-on effects, and so access for change would be limited to the system manager level.

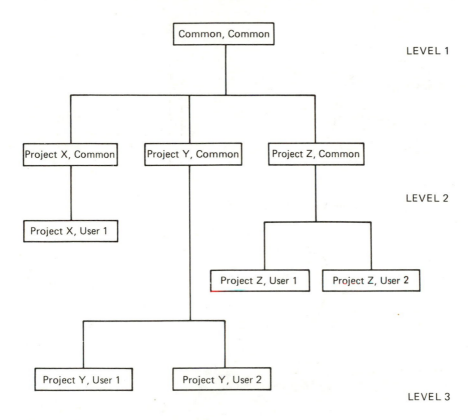

Figure 8.3   Hierarchy of storage libraries

Each project in the above structure would be self-contained and secure from access by unauthorised users in other projects.

How should parts be represented graphically?

This section is best illustrated with the aid of figure 8.4. It shows some of the possible methods of representing the same component within a computer.

With even the simplest draughting system there will be a multitude of alternative graphical representations of an engineering component—for example, to produce three unrelated orthogonal views and store each in a

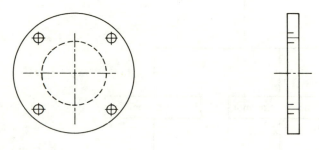

(a) Wireframe orthographic views

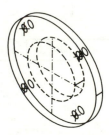

(b) Wireframe primitive

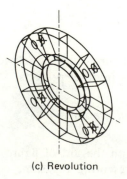

(c) Revolution

Figure 8.4   Comparison of wireframe and revolute type parts

library. To use the part, call the appropriate view out and use it. But what happens if a non-orthogonal view is required?

The only completely general method of representing a component available today is solid modelling. By its very nature this technique is unambiguous, but, unfortunately, because of the size of the computational tasks associated with it, solid modelling is not in widespread use. Even if it was, it may not be necessary completely to model every single engineering component in a library.

If a 'wireframe' model is used, there is a tradeoff to be made when establishing library parts. The component must contain enough geometry accurately to define boundaries, holes etc., but it must not be over complex otherwise removing hidden lines, for example, becomes time-consuming and tedious. The 'revolution' model (figure 8.4c) allows surfaces to be modelled, but at the cost of increased storage.

The final solution will always be a compromise. The representation chosen will contain sufficient detail to establish critical dimensions, but it will be primitive enough to be of general use, even if this does mean manually adding lines of sight to the final orientation of the part.

Is there sufficient disc storage available?

The answer to this question is almost certainly 'no', although with disc capacity increasing the potential size of 'on-line' storage is increasing. But, is it necessary to have immediate access to all live (that is, current production) parts? Again the answer is probably 'no', although at first thought it may be desirable.

The key to efficient use of disc storage is to hold on-line only those items being worked upon (components or assemblies), plus items which may be of use in more than one instance. All other items should be removed to the archive, and carefully documented so that retrieval presents no great time delay.

As an alternative to creating and holding individual components on the system, the concept of parametric modelling may provide significant benefits. *Parametric modelling* is the ability to design an item (such as a bolt) without fixing the basic dimensions (for example, thread size, diameter, length etc.). These dimensions can be varied and the system instructed to make one to the given sizes using a program. The values can be changed and a different version can be made quickly.

Such an approach is ideal for products which can be categorised by 'family of parts' relationships. The benefits of it are an initially reduced storage requirement—not every variant needs to be built and stored in a library. Against this must be weighed the effort needed to produce the original parametric program, and also the time taken for the system to create the item each time. Both of these may not be insignificant.

8.4.2 System Customisation

One of the major attractions of modern CAD/CAM systems is the high degree of customisation which can be achieved by writing special-purpose programs for standard operations. This is essential in order to maximise the throughput, and so provide the optimum return on the investment.

It is the responsibility of the project team to identify the priority areas to receive customisation. In the case study this was identified as the NC punch tape production, and it provides a good illustration of the quantifiable effect of customisation.

The NC tape preparation received high priority because of the need to supply the newly installed punching machine with control tapes (a key part of the justification, see section 8.2.2).

The basic software provided by Applicon (or any other vendor for that matter) for both geometric construction and control tape production was very general in nature. The operator was free to select his own conventions and standards, with the result that there was little consistency between operators and a great deal of typing (inputting of names) was required.

A realistic estimate of the operators' effectiveness at this point showed the computer-generated tape taking about four times as long to prepare as the manual methods. This needs to be qualified, since the computer-generated time included geometric construction that could be guaranteed to be accurate, while the manual method was prone to errors.

There are two phases to the tuning or customisation of the system—identification of the optimum method and development of a standard convention to cause this method to be used. A study of a large selection of sheet metal drawings, plus discussion with the operators, produced a list of shapes which were frequently needed:

(1) 45° chamfered corners;
(2) notches, both corners and 'vees';
(3) obrounds (oval holes);
(4) hexagons, dimensioned across corners and across flats;
(5) tapped holes (and shafts);
(6) keyholes;
(7) holes on a pitch circle diameter.

The standard userware could be used to produce these shapes, but it often required three or four separate commands to achieve the desired result. So some special-purpose user programs (called *macros*) were created to produce these shapes efficiently, and to allow their positioning using a pre-set datum. In this way sheet metal drawings could be rapidly and accurately entered into the CAD/CAM system.

The major portion of the NC punch tape production concerned instruct-

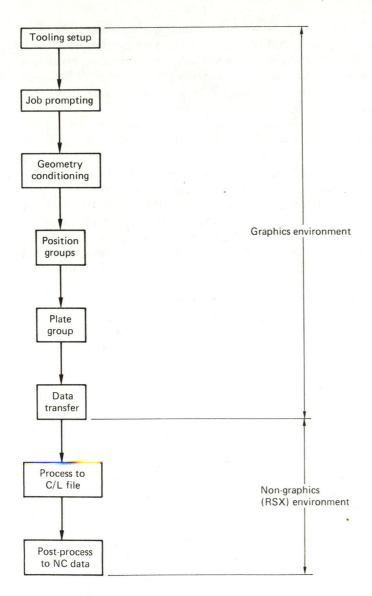

Figure 8.5 Flowchart for Numerically Controlled (NC) punch

ing the software whether to punch along the inside or outside of a closed boundary, and in which order to punch the groups of holes. Aspects of the process, such as tool path optimisation and nesting of the components on a sheet, were handled automatically by the post-processor for the machine tool itself.

The key to streamlining the whole operation was to adopt a standard convention for both names and colour/level usage on the system. Colour in fact turned out to be the most powerful aid to the operator, because it became possible to tell the state of a boundary (that is, whether inside or outside etc.) from its colour. Similarly areas of punch exclusion were identified to the operator by a change of colour.

A sequence of programmed function buttons was provided for the operator to enforce the chosen standard. The sequence is illustrated in figure 8.5. Tape creation for most components simply required a sequence of several button pushes, with appropriate geometry selections in between. Default settings for names, material, speeds and start-up conditions were built into the programmed buttons. The only typing necessary was the component identity and the revision number for the paper tape output.

Since the exercise was now highly automated, operators chose not to depart from the use of programmed buttons, because to do so represented more work. However, certain jobs did require non-routine operations—for example, to stop the machine and remove large slugs of metal. These could still be actioned using the standard Applicon features.

The customisation process took nearly five weeks, and to measure its effectiveness a timed trial was used with two opposite extremes of component complexity: a simple part with no more than 10–12 holes and a complex instrument casing with nearly 120 holes, including several irregularly shaped holes. Each part was drawn up and checked on the CAD system prior to the start.

The simple part was processed completely in 40 minutes; this compared with a manual estimate of 30 minutes. The program length required was also similar, with the automated process being only 5 per cent greated than the manual method. However, the automated process also had built-in checks that did not occur with the manual method. The advantage of the automated process was particularly pronounced for the complex part. The computer produced a tape in 3 hours compared with the manual estimate of 4, and the tape length was 25 per cent shorter than the manual program.

The improvement in productivity from the computer method was almost entirely due to the use of the new macros and the level standard, although the operator's skill had also improved somewhat. The result with the complex part was not surprising, but it was particularly satisfying to see the punch package deal with simple parts effectively, because this meant that the package was productive across the whole spectrum of the company work.

Other aspects of the system customisation which were also necessary included parts list production and training. Training was included here because unless further training is purchased from the vendor, it must be provided for in-house. This gives the opportunity to modify the training content to reflect its use in company, and to include any customisation already in place.

8.4.3 Utilisation and System Management

From the first day of the installation there were some essential procedures to be carried out.

The first concerns implementing a data back-up procedure to ensure the safety of any work carried out on the computer. This is to guard against both accidental deletion by the users, as well as catastrophic failures such as a disc head crash which could ruin a whole disc. The back-up consists of copying work from the disc on to a magnetic tape, which is then held in a fire-proof safe remote from the installation.

The nature of the back-up will vary from installation to installation, but typically it will involve a combination of incremental daily (that is, simply work that has changed that day) and full disc copies. The chosen method should ensure that no matter how serious a problem occurs, no more than the current day's work is lost from the system.

The second on-going task which falls to the system manager is to measure the system utilisation. This is important for several reasons. Firstly to show how frequently the system is being used (and if not why not?); it is normal to express this utilisation as a percentage of the available uptime of the system.

Secondly it provides a means of assessing how the implementation is progressing—which trainees have continued to use the system following training, and for how long. This information can then be reported back to the project team and higher management.

The choice of *System Manager* is quite important for he must possess some special skills, many of which are similar to those of a Project Manager (Chapter 3, section 3.4). Firstly, he must have the respect of the operators, not only for his system expertise, but just as importantly for his comprehensive understanding of the company's basic engineering techniques. A pure computer specialist is not the ideal choice for such a role, because he may lack the design knowledge to communicate with the operators on equal terms.

Secondly, in addition to this downwards compatibility, the System Manager must of course have the confidence of his superiors, who cannot expect to have anything like the level of detailed system knowledge that he possesses. As the last line of management, his advice directly affects the performance of the CAD system.

The system manager must also maintain co-ordination across the users of the system. Keeping the terminals all together in one area makes this one of the easier tasks. The transfer of information between the operators becomes frequent, and although not all of this relates to CAD/CAM, as the expertise of certain operators grows they are able to ease the demand made on the supervisor. Coupling this with the need to keep the workstations together during the training period, there is a good case for keeping them together unless there are extenuating circumstances.

8.5 CONCLUSIONS

8.5.1 Capital Justification

Various alternative methods for the capital justification of a CAD/CAM system were explored earlier. However, the installation of a CAD/CAM system provides a whole host of intangible (fiscally difficult to quantify) benefits which should not be ignored. This is especially true if the fiscal case is not particularly strong. The intangible benefits include:

(1) an ability to assess a wide range of alternatives before developing a design;
(2) a more rapid realisation of design;
(3) a more critical appraisal of designs using computer packages;
(4) a central store of all design information;
(5) an ability to generate CNC machine tapes without redrawing and using the central store of design information;
(6) an ability to test out tool paths, positioning of jigs and clamps;
(7) the simplification of modifications to drawings;
(8) access to a design store to assist in tool production;
(9) an ability to design systems using standard component designs and to optimise the three-dimensional configuration;
(10) the production of exploded drawings, possibly in colour for publicity literature;
(11) access to design information for inspection and quality control;
(12) an enhanced company image to the customers and sub-contractors.

Any or all of the above points should be used to add weight to the capital justification.

8.5.2 Project Evaluation

The implementation of a CAD/CAM system provides an interesting challenge in engineering management. It represents a cross between traditional short-term manufacturing-type projects (such as the introduction of a new process or technique) and the longer-term projects associated with civil engineering. It has similarities to both, but also major differences. The ideal project team should have experience of both types of project.

At regular intervals, the progress of the project to date should be reviewed. Topics for the agenda should include training, workload and system development. It is essential that if the project begins to lose momentum, the cause is quickly identified and action taken (Chapter 2, section 2.3.3).

There are a number of useful techniques available to measure the effectiveness of the implementation:

(1) On a regular basis (for instance, every two weeks) assess the competence of all the users.

Use a sliding scale with categories such as 'fully competent', 'requires minimal supervision', 'requires constant supervision', 'not capable of productive use' etc. This should provide pointers as to the development of trainees, as well as where to concentrate effort. It may well also indicate which trainees are unlikely to make competent users, and who therefore should be replaced by other more likely candidates.

(2) There will undoubtedly be pressure to measure the CAD/CAM productivity gain by comparing it with manual methods. If this is the case, ensure it is under controlled conditions. The only true way to do this is actually to duplicate work. The more usual way to do it, however, is to carry out the task on the CAD system and estimate the manual equivalent.

Under these conditions, ensure that several manual estimates are used—people invariably under-estimate manual times, so temper these with actual experience. The choice of the actual task can be used to great effect for 'time trials'. Ensure that the piece is well suited to CAD (some types of work, like piping or wiring, are still notoriously difficult to produce efficiently on CAD). Alternatively, use a component which is similar to one already in the database; it can be used as a starting point with excellent results.

In short, for direct comparison, the work can be manipulated to give whatever results are required! It should also be noted that the CAD system provides most of its benefits downstream of the design process (for example, in producing parts lists). For a full comparison, the complete design/manufacture/modification cycle will shown the CAD/CAM system in a more realistic manner than just the design process alone.

A typical comparison of relative times for different activities might take the form shown in table 8.3

The significance of these relationships is that of all the processes, the design and detailing offer the least potential gain. Indeed it is felt that a figure of 1:1 productivity from the design process should not be considered as disappointing. To offset this, massive productivity gains are available from say Bill of Materials generation, modifications and the design of 'same as excepts'. Hence when the figures are summed across the whole spectrum, the CAD system can give around the 3:1 productivity gain often quoted by the vendors.

It is an important aspect of the implementation to ensure that the high productivity areas are brought on-stream as soon as possible. This is especially true if the design and detailing are showing somewhat less than 1:1 productivity.

8.5.3 Project Management

It is necessary to quantify the success of an installation as early as possible. If a great deal of effort has been expended on customising the system, then it must be thoroughly tested for effectiveness. The ideal situation for a

Table 8.3 Comparison of Activity Times With and Without CAD

Activity	Design	Detailing	BOM	NC tape	Modification	Design of 'Same as Excepts'
Manual	10	10	10	10	10	10
CAD	8–10	5–8	1–2	4–5	4–6	1–4

CAD/CAM installation is the vehicle of a new (or substantial redesign) project. This will provide the necessary ingredients to ensure that momentum builds up and is maintained throughout the installation, for two reasons. Firstly, the project requires completion against a deadline, and this will ensure all necessary co-operation and assistance. Secondly, there is an incentive to the operators to complete the task, as opposed to just copying existing work.

The only potential risk concerns the project's timescale. This must be realistic in CAD terms. Otherwise when problems occur, as they inevitably will, there will be insufficient time to correct them without causing the project to slip badly behind its schedule. This simply creates more pressure on the operators and system supervisor, probably causing the project to slip further back. Under these circumstances, careful management is required to prevent the project being withdrawn altogether from the CAD system. Such a course of action causes a loss of credibility for all concerned.

In conclusion, a CAD/CAM project needs care in all its stages: feasibility, justification, selection and implementation. This is of course the case in any project, but since CAD/CAM is an example of modern technology it has its own additional constraints on a project such as that discussed here.

8.6 DISCUSSION QUESTIONS

1. What are the circumstances that favour the adoption of a CAD/CAM system?
2. What is benchmarking, and how can it be used to complement the other techniques needed for system selection?
3. Illustrate why it is difficult to produce a traditional financial justification for a CAD/CAM system.
4. How should the users of a CAD/CAM system be selected and trained?

5. Why does it take 12–18 months for a CAD/CAM system to produce an improvement in productivity, and where are these improvements most likely to be found?
6. How should the database on a CAD/CAM system be designed and built up?
7. What are the attributes required of a CAD/CAM system manager, and what is his role?

Chapter 9
A Manufacturing Case Study

9.1 INTRODUCTION AND PRODUCT DESCRIPTION

This chapter uses one medium-sized project carried out within an engineering company as a case study. The company is not mentioned by name but it manufactures machines used in producing a food product. The underlying theme is quality, in that the aim of completing the project was to improve the quality of the machines. The intention here is to pick out the key stages, or prerequisites for success, and to draw these key points together at the end of the chapter, to allow their use as guidelines for the planning and/or implementation of other projects.

The machines are linked together to constitute a process that is sold to customers who wish to produce a consumer product. At certain stages the materials are moving through the machines at up to 70 m/s, so they are very high-speed machines. As with all process machines, it should be noted that the control of the material through the process is critical, and if at any time precise control is lost or even diminished in any area, then that area will be a potential problem area throughout the life of the machine. This means that the machines have to be well engineered, and that high standards of quality in manufacture have to be maintained. In fact, as the company has continued to develop newer and faster machines, the reliability of these machines has become increasingly dependent on high-quality standards.

It is worth detailing the nature of the machines before going into the project in any depth, as this information is necessary to understand why the project was undertaken, and why it was so detailed.

There are three main machines produced at the site in question; these are linked together in the way shown by figure 9.1. The first has a hopper, which is loaded with an expensive material, which is then presented in a suitable form to the main part of the first machine (the Producer). In the

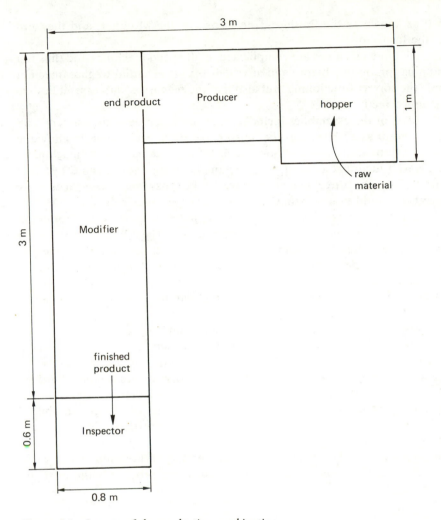

Figure 9.1   Layout of the production combination

Producer it is combined with another material to produce what may be called the end product. However, at the opposite end of the machine to the hopper is a drive shaft, which is used to drive a second machine. In 95 per cent of the cases the second machine (the Modifier) is linked to the first machine (or alternatively to a competitor's machine) to convert its end product into a more refined item. It is unusual for the first machine to produce an end product. With the second machine fitted, the end product of the first machine is fed directly into the second machine, where an additional section of another material is added to complete the finished product. The third machine is also optional, and it inspects the finished product as it leaves the second machine. From now on, the first machine will be known as the Producer, the second as the Modifier, and the third as the Inspector.

All three machines are complicated, high-speed machines and thus test running (preferably before and after delivery) is essential to guarantee not only the correct functioning, but also that the machines' sub-assemblies are set up correctly.

Most of the assemblies included have some degree of adjustment, and the adjacent assemblies must be timed correctly to each other, to allow the product to be transferred without damage from one part of the machine to the next functional part of the machine. During the machine test and installation, the three machines have to be positioned very accurately together, in addition to being timed together very precisely.

Because of the high cost of the primary material used by the Producers, they alone are tested using the correct material. The Modifiers are linked to test rigs which present them with a similar product but made from cheaper material. The cheap material is important, as the Modifiers take a lot of test running to set up and time correctly. When this has been achieved, the Inspector is linked to the Modifier and the two machines are run on a test rig until both are running correctly.

The function of these machines means that they can best be defined as a process, but a process making products which vary in size and shape considerably. This results in the three machines having a large number of variable conditions, for producing the different sizes of end product. Furthermore, because of variations in the quality of the materials used in them by different countries, there is a very large degree of design variability, which affects the link-up conditions from the Producer to the Modifier, and from the Modifier to the Inspector. It is worth noting that the Inspector's set-up conditions also affect the Producer, because of the complex electronic controls that are used to monitor and control the product being made by these machines.

9.2 PROJECT BACKGROUND

Over 90 per cent of all the machines that are produced are exported, with a large proportion going to the USA. Major factory reorganisations in the USA can mean that, at any one time, up to 75 per cent of each month's production can be for one US customer. A complete month's production programme is usually about 20 machines of each type.

The complexity of these machines means that nearly all are installed by the manufacturer's own field engineers. The exception is the USA where the factories have installation teams trained by the manufacturer's US subsidiary, and these teams operate permanently in the large US factories, because of the continual installation of new machines on a large scale.

This project was carried out at a time when, for an extended period, a large proportion of each month's factory output consisted of machines being produced for one specific customer in the USA. However, with the order, came a complaint that the US installation teams in their factory were having difficulty linking the combinations, and running them together. The new order specified that all combinations must be linked during the machine test, before the machines were despatched to the USA. The first two groups of machines (that is a Producer, Modifier and Inspector from now on known as a combination) were duly linked and tested in-house, a process which took two weeks, and represented a serious bottleneck and threat to the delivery dates.

The project detailed here was to reduce the link-up test time from two weeks, to two days, and the following terms of reference were issued.

9.3 THE PROJECT TERMS OF REFERENCE AND THE TESTING SEQUENCE

9.3.1 The Terms of Reference

Terms of Reference

'Because of the increased load on the Assembly Test Section caused by the new American Test Requirement it has become necessary to reduce the amount of time required to link up and test a Producer, Modifier and Inspector combination.

The project will be undertaken by a Project Engineer who will report directly to the Production Manager.

The objectives of the project are:

(a) To analyse current test procedures and identify deficiencies, and
(b) to outline revised test procedures aimed at reducing the test time of a linked combination to 2 days.

This will involve working with Assembly Testers and the Project Engineer will be responsible through the Test Foreman to the Assembly Manager while working in that area.

Signed . . . .   Copies to: The Project Engineer
Production Manager   Assembly Manager

It should be noted that this was the first time a university graduate of any sort had been used on the shop-floor in this company. Also, although it might be reasonable to assume there had been considerable discussions between the Production Manager and Assembly Manager, the Production Manager did not introduce the Project Engineer to the Assembly Manager, but issued the copies of the Terms of Reference and despatched the Project Engineer to the Assembly Shop to carry out the work.

The Terms of Reference have three shortcomings from a Project Management Standpoint (section 3.3):

(1) The Test Foremen did not receive copies. If the Assembly Manager was known by his Production Manager to be a good communicator, this might not have been necessary, but as it turned out he was not. The result of this was some degree of alienation with the Test Foremen as they could not explain to their Testers who the Project Engineer was, what he was doing (in detail), why he was doing it, how long it would take, what was needed regarding testers involvement, and whether it involved other sections in the Assembly Shop apart from the test area.

(2) The Union Representatives were told nothing at all about the arrival of the Project Engineer in Assembly or what he was doing there.

(3) No timescale was put on the project regarding completion. It could be argued that this was due to little being known about the project content, but there was no mention in the Terms of Reference of an initial fact-finding period after which target dates might be set. It may have been felt that these might be unrealistic and would have to be revised continually. An alternative would have been to set dates for a series of progress meetings which might, for example, be monthly. In this particular case this could have been a critical omission, as it was the first project carried out by this Project Engineer in the company after graduating. Although the Production Manager had experience of work from the Project Engineer's industrial training periods, the fact that it was the first major project carried out on the Shop Floor by a graduate meant that scheduled meetings should have been organised. Additionally, the Terms of Reference should have specified who was to be involved in such meetings, for example, people such as the Production Manager, Assembly Manager and the Project Engineer, and whether the Test Foremen would also be involved.

The line of responsibility was clearly outlined in the Terms of Reference and that was good. The explanation of why there was the sudden need for the project to be carried out was also good. All the more reason to have sent copies to the Test Foreman and Union Representatives!

Carrying out a project of this nature with only a small amount of knowledge regarding the operation of the machines under running conditions, and knowing nothing of test methods, meant that it was not possible to forecast the kind of detail it would be necessary to record about test procedures and the occurrence of faults. As a consequence there were two primary objectives. The first was to establish an overview of the procedures involved in testing linked combinations of machines, and the second was to establish the best method of monitoring the testing of the individual machines, the testing of combinations and the logging of the difficulties and faults.

9.3.2 The Testing Sequence

Through discussion with the two test foremen, it was established that the Modifier was set up (a process which took approximately one week) before it was linked to a test rig, in order to run the dummy product made from cheap materials through it. This was to check the correct functioning of the machines and the quality of the product. Next, if an Inspector had been included in the customer's order this was linked alongside, having already been fairly crudely tested during the previous week on its own simple test rig. When both were functioning correctly, they were moved to the bay where the Producers were made. The Producer to be linked with the Modifier would have been set up and tested at the same time, but separately from the Modifier. The machines were linked, and again if all runs well, or after any faults have been sorted out, the Inspector was added and the complete combination was run under full operating conditions, while the final checks were carried out. Having established this basic outline, it was suggested that since a combination had just been linked together, it would be a good idea to follow the test through to the final passing-off of the machine by Quality Control. (The fairly unusual line of authority should be noted here. The Machine Test Foremen and their operators reported to the Assembly Manager, as did the Sub-Assembly Build Foremen and the Main Machine Build Foremen. It would perhaps have been more usual for the Test Personnel to report to the Quality Manager, or one of his members of staff. The deficiency with the system used in this example was that the Assembly Manager was controlled by shipping dates. The Assembly Manager was therefore under continual pressure to compromise the product quality, by shortening the test procedures on machines held up by faults, in order that the shipping dates could be met.)

It is worth deviating briefly here to explain the method used for logging faults, and why it was chosen in preference to other methods.

9.4 METHODS OF LOGGING INFORMATION

9.4.1 The Philosophy behind the Testing

It was decided that any combinations being tested, whether for the USA or another customer, would reveal valuable data. Although the Producers, Modifiers and Inspectors were tested separately and monitored in this investigation, if they had a different link-up condition they would not be monitored as a combination. This decision was made on the basis that not all the machines produced could be monitored, and not even all the linked combinations could have the testing of the individual machines covered in detail. So within these terms of reference, as many as possible link-up tests were monitored in the time available, and any gaps between the link-up tests were used: firstly to study the individual machine tests, and later to look at the organisational procedures.

The other reason why it was decided to concentrate the immediate effort on the monitoring of the complete combinations was that this was the first stage at which the machines ran for long periods, and this was where the functional problems were more likely to occur. The fact that the faults were found at such a late stage appeared unimportant, as it was assumed that by using simple logic it would be easy to find the source of the problem.

For example, if a machine malfunctions this can be traced to a sub-assembly and hence to a component or several components. These may be checked against a parts list to see if the correct parts have been used and, if not, the fitter responsible for checking the parts to the parts list is at fault. If all the parts are correct, they will be checked against drawings to find which deviates from its specification. When the non-conforming items are found, if they have been manufactured in-house, the implication is that the inspection systems should have identified them as non-conforming, or if they are bought-out items the Goods Inward Inspection should have tested or measured and rejected the parts. In this way deficiencies should be traced quickly.

9.4.2 The Method of Logging Fault Data

The next consideration was the method by which information was to be logged. Although the method used is best described as a diary approach, it is different from that which is normally described as a diary technique, and that described by BS 3138.

The disadvantages of the standard diary technique are associated with inaccuracies due to other people monitoring what goes on rather than the researcher. The associated advantage is that more work can be covered by a team of researchers. However, in this case, because the Project Engineer was doing the monitoring himself, these inaccuracies did not occur. It must also be considered that initially the detail in which events were to be monitored was not known, thus the flexibility of the diary approach was ideal to monitor anything and everything that occurred. This would continue until a reasonable level of detail could be established, which would reveal the number, type and nature of the machine faults, and the amount of lost time attributable to the different causes.

Furthermore, the diary method was seen to suit a situation in which little time was available to go into detail at the time the faults occurred. It was therefore necessary to use a diary to make short notes from which a more detailed report could later be made.

9.4.3 Fault Logging in Practice

The result of the diary approach was that no specific shorthand or abbreviation was necessary to log the information required. A dictating machine might have helped, but background noise made this impractical. Provided each set of notes was extended into a report of reasonable detail the same day, little or no detail was lost as a result of forgetfulness.

The first combination was tested over a period of about a week. This consisted of intermittent running, interspersed with fault finding and fine tuning. The faults were logged as they occurred, with a date and time, to build up a pattern of at which stage a component failed or a malfunction was detected. The main objective of the Project Engineer at this early stage was to be amenable to the Tester carrying out the test, not to upset or obstruct him in any way, and to lend asistance where possible. Initially this was not so much in a technical way, but by helping to clear up, by loading materials into the machine and by sweeping up when the spilled materials became a nuisance on the floor around the machine. The willingness to help (without insistence) is the key to breaking the ice in situations such as this. A willingness to communicate freely about the work being undertaken to personnel, whether skilled or unskilled, is also important. Common interests such as football results, TV soap operas and hobbies will feature to a major extent in the early stages of the conversation. All these things help an outsider to become accepted on the shopfloor, by finding conversation topics of mutual interest.

The most significant occurrence during this first week was a Union Meeting involving all the Testers on the second day, to discuss why the Project Engineer was working with them. During this meeting the level of apprehension among the Testers increased appreciably, as none had been told in detail by the Assembly Manager why the Project Engineer was working with them.

This situation was diffused very quickly. As soon as the Project Engineer realised that no-one had been briefed adequately, he gave his copy of the Terms of Reference to one of the Test Foremen, to read out aloud at the meeting and show the Union Representatives. This, with an offer to give them their own copy, was sufficient to placate them and work was resumed. It should be noted however, that this was a very benign union, and had there been militant elements present a full dispute might have resulted. This underlines the critical importance of communication and consultation before a project commences, as advocated in section 3.3.2.

During the successive tests, a number of linked combinations followed with much the same pattern as the first combination—a few malfunctions and intermittent running for fine tuning. All this was logged as each event occurred. In between these tests, the setting up of either the Producers or the Modifiers on their test rigs was monitored in detail, to establish the exact procedures. Most of the personnel were not strict Union Members, and the Project Engineer shared some of the setting and testing tasks under the supervision of the Tester. This enhanced the Project Engineer's understanding of the machines' functions, and helped to demonstrate some degree of technical competence, which again helped him to be accepted by some of the more reticent test personnel.

It would be pointless to go into the technical details, but there were two major outcomes from this work, with a number of peripheral events.

The two major outcomes were discrete, and were the result of an outsider looking objectively at the test procedures without becoming bogged down in detail. This is the true value of introducing an outsider to look objectively at a situation. Firstly the Quality Control management systems needed to be improved, and secondly the need for Quality Assurance through the testing of individual components and systems was highlighted. Additionally, it was shown that the Electronic Testers were not being trained properly.

9.5 QUALITY CONTROL

9.5.1 Introduction

The log book showed that there was no pattern of set faults but a random succession of mainly technical faults; these faults took the form of the malfunction of a 'bought-out' item or the lack of conformance of an item manufactured in-house.

The lack of a pattern indicated that no specific components or assemblies were at fault. However, the similarity in nature of the faults (that is, non-conformance of manufactured items or failure of bought-out items), and the frequency with which these faults occurred indicated, as was suggested earlier, that they were in fact related. Although they all had discrete technical solutions, their continued occurrence indicated an organisational problem. This problem was that the defective purchased items were not being identified before being fitted to a machine, and that the non-conforming manufactured items were passing undetected through the machine shop inspection systems.

The implications for an organisation having inadequate inspection in the Manufacturing and Goods Inwards Departments are far reaching. The importance of quality control in engineering projects was mentioned in

Chapter 2, section 2.3.2. Furthermore if the conformance of bought-in items is not checked, the purchaser's statutory rights can be affected—see Chapter 4, sections 4.4.4 and 4.6.2. The more thorough the early testing of the assemblies and components, then the easier the testing of the whole machine combinations. This is because if two assemblies are thoroughly tested individually, and found to be correct, but fail when linked, then the confidence in the individual assemblies allows the test engineer to look immediately at the link-up area alone. If no such confidence exists, the failure could be anywhere in either of the two assemblies, resulting in considerable time being necessary to pinpoint the cause. This is demonstrated by figure 9.2, which shows the way that poor quality control leads to cost escalation. The next section illustrates and emphasises this point.

9.5.2 Automatic Test Equipment

The electronic controls on the machines are extensive and complex. The general checks carried out on the machines, prior to the power being switched on, took 2–3 days before the test running could begin. In addition, the dependence of the Modifier and Inspector on the main electronic control panel of the Producer made fault finding very difficult and time-consuming, once the units had all been linked together. The control cabinets were built up in the following way.

All the printed circuit boards, wire-wraps and wiring looms were sub-contracted to a separate division called SEE. Their production was scheduled to the main site's factory programme, and the assemblies were held in an electrical store with the components and items that were bought-in directly. At SEE all these components were tested using *Automatic Test Equipment.* This equipment was microprocessor based, and functioned by putting signals into the printed circuit boards and measuring the outputs. An example that was known to be correct was used to teach the test equipment how a correct item should react, after which any deviation in the response was reported. Additionally, it could be used to test the wire-wraps or looms as a simple point-to-point or continuity tester.

The first thing to be highlighted in this project was that the wire-wraps were being damaged in transit. This resulted in items which were known to be correct at test arriving in a defective condition. This was corrected by a change in the handling and transport procedures.

The main control cabinet for all three machines was in the Producer, and the control cabinet consisted of a power column, the control and protection gear such as thermistors for motors, the product monitoring electronics and the control electronics. This was made up from the items supplied by SEE, but assembled together at the main site. The main control cabinet was then plugged into a function tester, which had been made

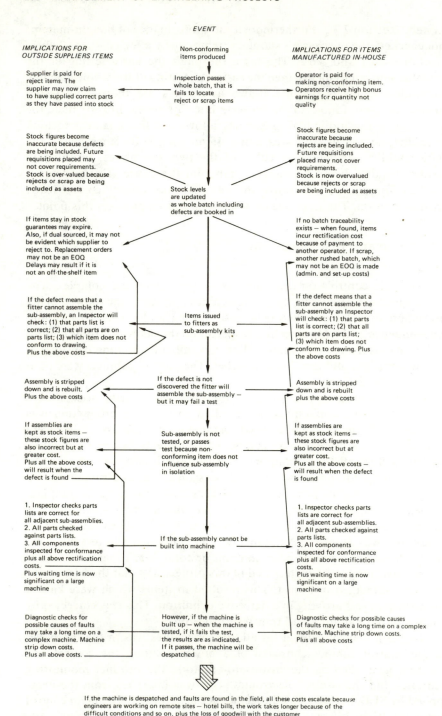

EVENT

IMPLICATIONS FOR
OUTSIDE SUPPLIERS ITEMS

Non-conforming
items produced

IMPLICATIONS FOR ITEMS
MANUFACTURED IN-HOUSE

Supplier is paid for
reject items. The
supplier may now claim
to have supplied correct parts
as they have passed into stock

Inspection passes
whole batch, that is
fails to locate
reject or scrap items

Operator is paid for
making non-conforming item.
Operators receive high bonus
earnings for quantity not
quality

Stock figures become
inaccurate because defects
are being included. Future
requisitions placed may
not cover requirements.
Stock is over-valued because
rejects or scrap are being
included as assets

Stock levels
are updated
as whole batch including
defects are booked in

Stock figures become
inaccurate because
rejects are being included.
Future requisitions
placed may not cover
requirements.
Stock is now overvalued
because rejects or scrap
are being included as assets

If items stay in stock
guarantees may expire.
Also, if dual sourced, it may not
be evident which supplier to
reject to. Replacement orders
may not be an EOQ
Delays may result if it is
not an off-the-shelf item

If no batch traceability
exists — when found, items
incur rectification cost
because of payment to
another operator. If scrap,
another rushed batch, which
may not be an EOQ is made
(admin. and set-up costs)

If the defect means that a
fitter cannot assemble the
sub-assembly, an Inspector will
check: (1) that parts list is
correct; (2) all parts are on
parts list; (3) which item does not
conform to drawing.
Plus the above costs

Items issued
to fitters as
sub-assembly kits

If the defect means that a
fitter cannot assemble the
sub-assembly an Inspector
will check: (1) that parts
list is correct; (2) that all
parts are on parts list;
(3) which item does not
conform to drawing. Plus
the above costs

Assembly is stripped
down and is rebuilt.
Plus the above costs

If the defect is not
discovered the fitter will
assemble the sub-assembly —
but it may fail a test

Assembly is stripped
down and is rebuilt
plus the above costs

If assemblies are
kept as stock items —
these stock figures are
also incorrect but at
greater cost.
Plus all the above costs,
will result when the
defect is found

Sub-assembly is not
tested, or passes
test because non-
conforming item does not
influence sub-assembly
in isolation

If assemblies are
kept as stock items —
these stock figures are
also incorrect but at
greater cost.
Plus all the above costs —
will result when the defect
is found

1. Inspector checks parts
lists are correct for
all adjacent sub-assemblies.
2. All parts checked
against parts lists.
3. All components
inspected for conformance
plus all above rectification
costs.
Plus waiting time is now
significant on a large
machine

If the sub-assembly cannot be
built into machine

1. Inspector checks parts
lists are correct for
all adjacent sub-assemblies.
2. All parts checked against
parts lists.
3. All components
inspected for conformance
plus all above rectification
costs.
Plus waiting time is now
significant on a large
machine

Diagnostic checks for
possible causes of faults
may take a long time on a
complex machine. Machine
strip down costs.
Plus all above costs.

However, if the machine is
built up — when the machine is
tested, if it fails the test,
the results are as indicated.
If it passes, the machine will be
despatched

Diagnostic checks for possible causes
of faults may take a long time on a complex
machine. Machine strip down costs.
Plus all above costs

If the machine is despatched and faults are found in the field, all these costs escalate because
engineers are working on remote sites — hotel bills, the work takes longer because of the
difficult conditions and so on, plus the loss of goodwill with the customer

Figure 9.2   The organisational implications of inadequate quality control

in-house. The function tester required an Electronics Tester to work through a test specification, throwing switches and checking the response. This task took about 1.5 days, and was unreliable. The problem was that if a tester had tested a series of cabinets, he might come back from lunch, for example, and start off on the wrong paragraph of the test specification, thus omitting to do some tests by accident. Also, the boring repetitive nature of the work tended to result in the testers overlooking some faults. This automatically nullifies the testing, because there can be no confidence in a partially tested assembly.

A piece of Automatic Test Equipment, which was capable of function testing control gear of this complexity, would have cost in the region of several hundred thousand pounds. The logical option is to build the cabinet with all the wire-wraps and looms, and then to do a complete continuity or point-to-point test. This will give a high level of confidence in all the wiring and connections, and would be sufficient, because there is a high level of confidence in all the printed circuit boards because of the tests at SEE. Therefore, if all the wiring and connections are correct, when conforming circuit boards are added the complete assembly should function correctly.

The capital proposal for some Automatic Test Equipment did not progress very far, because it was not a function tester. It is worth noting that this work came at a time when the company was changing from NOR-gate logic to microprocessor controls for the machines, and consequently several tailor-made function testers would have been required. Also, as has been stated, previous ones had been built in-house and there was little time for new ones to be modified or built. Automatic Test Equipment (ATE) capable of learning the characteristics of good assemblies in seconds seemed the obvious solution, particularly as a complete cabinet could then be tested in 10 minutes (including the connecting and disconnecting of all the plugs and the reading of a printout). The actual test duration was approximately 10 seconds.

A further reason for ATE was that a number of different types of controller were being introduced, and the ATE could learn the circuit functions from good products, and store all the test information for the many different circuits and models. In comparison, a function tester that was built in-house would be dedicated to only one product.

The other reason why the proposal did not receive ready acceptance was that it was not based on reducing staff levels. It took some time for the management to realise that taking less time by a reduction of 12 hours to 30 minutes was not the major benefit. The major benefit from a Quality Control aspect was that final fault finding would be expedited because there was a very high confidence level in the tested item being faultless. The point being made here is that when three electrically complex machines are linked together, to have a high level of confidence that the major control system is correct saves a significant amount of time when fault finding. It can be assumed to be correct and only investigated as a last resort, thus concentrating effort in a few easily tested areas.

These two major pieces of work—the tracing of the technical faults to organisational deficiencies and the explanation to management and supervision of the importance of having high confidence in the quality of components, then small assemblies, main assemblies and finally the complete products—marked the beginning of the move from Quality Control to Quality Assurance. Quality Control has traditionally meant the checking of finished items, while Quality Assurance is the building into a product of quality at every stage.

9.5.3 Progress Review

One of the most important aspects of project management is to fulfil the objectives of the project, as originally stated, or as revised during the execution. For this reason some project engineers start each and every day by reading through their Terms of Reference. This is by no means a waste of time or bad practice, as it concentrates their efforts on directly relevant issues. For this reason the Terms of Reference are repeated here to remind the reader of the original objective.

Terms of Reference

Because of the increased load on the Assembly Test Section caused by the new American Test Requirements it has become necessary to reduce the amount of time required to link up and test a Producer, Modifier and Inspector combination.

The project will be undertaken by a Project Engineer who will report directly to the Production Manager.

The objectives of the project are:

(a) To analyse the current test procedures and identify deficiencies, and
(b) to outline revised test procedures aimed at reducing the test time of a linked combination to 2 days.

This will involve working with Assembly Testers and the Project Engineer will be responsible through the Test Foreman to the Assembly Manager while working in that area.

Signed ....   Copies to: The Project Engineer
Production Manager   Assembly Manager

As can be seen, the progress to date has been concerned with instigating an organisational reform to prevent the faults in assemblies reaching the final test stage. These measures should have an equally beneficial effect on the tests of individual machines as on the linked combinations, because the emphasis is on keeping things simple and checking items at a stage when they are easily checked. By eliminating faults in this way, sometimes referred to as a '*Zero Defects Policy*', it is possible to reduce to an absolute minimum the chance of failure at test or in the field (the two times at which faults cost the most, as seen from figure 9.2).

9.6 TEST PROCEDURES

The next avenue to be explored might appear initially to be a deviation from the Terms of Reference, but as will be seen, it is just as relevant as the earlier work.

The investigation of the electrical/electronic test procedures called into question the method by which all the test staff were trained. This tended to be by standing beside an experienced tester and learning what had to be done, and to some extent how it was carried out. The complexity of the electronics in the machines meant that this was thoroughly unsatisfactory, as when a fault occurred the immediate reaction was to look at what was the previous cause of a similar problem on an earlier machine. As a result of this project, the Electronics Testers were sent on Service Engineer Training Courses, which taught the theory of how the machines were supposed to function. This led directly to a diagnostic approach to fault finding, rather than a random searching. It should be noted that with very complicated machines, the designers' idea of how the machine is intended to function is not always obvious. Added advantages were that testers met the customers' personnel during training, and learned of their problems. In addition, the testers felt someone was taking an interest in their careers and work. The immediate advantage of these courses was that the diagnostic fault finding methods cut the original electrical test time of 2–3 days by one-third. But as with the previous work, the main advantage was that after a thorough diagnostic check-out and test, there was much higher confidence in the machine functioning correctly. There were now fewer instances of any electrical faults or bugs on individual machines, and hence on the linked-up combinations.

The overall picture that emerged was that the project was affected by a tremendous number of external factors. To achieve a relatively simple objective meant affecting a number of different departments in various parts of the factory, each of which involved a fairly large piece of work.

Above have been mentioned a number of courses of action which, after investigating the situation, appeared the most likely to generate the appropriate results. There were a number of other factors which contributed to the successful completion of the project, some of which were as a result of the Project Engineer being there, and some of which were not. Consider the following brief examples.

A less significant organisational change was that, formerly, before the link-up tests, the Producers had been built and tested in one of the assembly bays and the Modifiers were built in a different bay and were run-up using only test rigs. Consequently the testers, who were generally specialised in one or other of the machines, were based in one or other of the bays. The need to transfer Modifiers into the Producer bay for the test meant that considerable amounts of time were being wasted fetching tools. In addition, several specialised tools were held on the Modifier test section for special settings, and these were now also being used during the link-up tests. As a result, it was recommended that these be duplicated, and the testers be issued with tool trolleys to avoid wasted time. A side effect was that the testers felt the management had recognised their work. Although this was only a relatively small time saving compared with the other savings, it was significant, and it would not have occurred had the Project Engineer not been there.

One of the main reasons why the tests began to take less time and the procedures were carried out more effectively was that formerly only two or three combinations had been linked per year. Whereas when the testers had to test many combinations per month they became much more accustomed to the procedure; that is, the test procedures became more polished and a rhythm developed in the test teams for these procedures. This would still have resulted in a major time saving, even if the Project Engineer had not been working in the test area.

It is worth remembering also that formerly all these link-up problems had been experienced by the company's Service Engineers during installation work, and the problems had been documented in their reports. However, they had had to overcome the problems when in customers' factories, and a problem solved carried little real importance with the managers at home. Indeed, since the managers were remote from the problems they did not necessarily appreciate how severe the problems were. As soon as these problems occurred on the home site, the designers and quality engineers were brought in to deal with them.

It should however, be noted that this in itself was not enough to get all the associated projects going. For example, the ATE project was never approved, although it had been firmly initiated by the Project Engineer. However, the Project Engineer was responsible for the introduction of extra tooling, air-lines, services and the tool trolleys. To summarise, some of the issues mentioned here needed a Senior Manager associated with the project to push the situations and deficiencies to the attention of the designers, the quality staff and the management, so that the necessary action was given the appropriate commitment and drive.

9.7 QUALITY CONTROL OF EXTRAORDINARY ITEMS

There was one more example of how the drive of an individual committed to the project achieved a great deal. It came unexpectedly in the form of some modified circuit boards which arrived from the US subsidiary. The accompanying instructions were to take out two of the existing boards after the final machine link-up test, and to replace them with one of each of the two new ones delivered from the USA. In other words, the customer had collaborated with the US subsidiary, and the subsidiary had modified two types of circuit board.

To change components on a machine which has had a final test, and is not to be run again until it is in a customer's factory, is a totally unacceptable practice from a Quality standpoint. This was more critical as special test procedures were being used to ensure the good quality of these machines in particular and, moreover, the inclusion of printed circuit boards which could not be tested on site (and had been manufactured elsewhere) meant that the quality of the machines could at this stage be compromised by 'foreign' components.

It could be argued that reading the Terms of Reference shows no mention of an investigation into this 'deviation'. However, an interpretation of the objective of the project was to improve final product quality, and this could be jeopardised by not investigating the current situation. The Project Engineer could thus be forgiven for showing some initiative and investigating these US instructions.

The Project Engineer started by asking the Supervisor in charge of all the electrical/electronic build and test work, whether he intended to obey the instruction. The reply was yes, because the US instructions told him to Next the Chief Electrical/Electronic Design Engineer was asked whether he agreed to machines being despatched untested, and with unknown conditions on them. It was also pointed out to him that this was a time when the machine quality was of the utmost importance, and the company was risking goodwill with the customer and its quality reputation by obeying the instruction from the USA. In spite of all these facts he would not make a decision one way or the other. It was known that the new printed circuit boards (PCBs or boards) could not be tested on site, but by visual inspection it could be seen that three of the new boards had solder splashes on the back that joined several of the tracks together. Unless this had happened after they had been tested the PCBs could not have been tested. If they had been tested, they had not been carefully stored afterwards. Either way, these faults showed that the new PCBs could not just be fitted without testing.

The next step was to test a Modifier until it ran correctly (that is, until it was known that no faults were present) and then to fit two of the new PCBs. In the following test run, one of the machine's functions intermittently stopped and started, causing the manufactured product to be inconsistent.

The machine seemed likely to be functioning incorrectly, so the PCB affecting that function (which was one of the new ones) was replaced by another new one. This time the machine functioned as it would normally have done.

Next a Telex requesting more information was sent to the engineer in the USA who had initiated the change. Through a telephone call later that day he explained the following:

(1) The function of the machine should not be altered in any way, but that the new format of the PCB allowed an extra link to be added which slightly altered one function thereafter—a facility requested by the customer.

(2) In response to a question about his instruction not to test with the new boards in place, he replied that it had not been an instruction, but some advice to say that the PCBs had all been pre-tested in the USA, and as a result of that the machines would not need testing again with the PCBs in place. This was in case some of the machines to have the new PCBs fitted were already in the Despatch Area or crated up.

When told that some of the PCBs were defective he was first annoyed that his instructions to his own staff in the USA had not been carried out. Then, as he realised the consequences for the English company of obeying his instructions and fitting defective PCBs, he was apologetic for inadequately supervising his staff on the testing and subsequent care of the boards. In addition, he also apologised for not supplying more information about the effects of fitting the new PCBs.

This still left one problem to be solved—three combinations of machines to receive the new PCBs had indeed left the test area and would miss their delivery dates if they had to be retested at this late stage. The answer was to test the boards before fitting them, and this was achieved by using a machine still in the test area that was known to run correctly. This machine was fitted in turn with three sets of the new PCBs and, since the machine ran correctly with each set in turn, this indicated that all three sets were known to be correct. These tested boards were then used to replace the old format of PCB in the machines in the Despatch Area.

As may be seen, from a quality point of view, some fundamental bad practices could have lost the company a very great deal of goodwill and harmed its reputation, at a time when it was taking considerable steps to improve its 'Quality Image'. At the same time, it should be noted that the Design Team in the US subsidiary had not been instructed in the fundamental importance of simple quality control rules, such as NEVER tampering with tested products which are ready for despatch. The critical importance of testing and the subsequent care of components going straight into a customer's plant were underlined. But more importantly than both of these lessons, the fundamental importance of adequate supervision of staff at all times was emphasised. This also comes back to communication as, if the designer had explained and emphasised to his staff the importance of the new PCBs all being correct, they might have been a little less careless with them.

It should be underlined that this particular piece of work was carried out under the continual pressure of delivery dates, in the form of flights already booked to the USA. The reason why the designer was not consulted sooner was the time difference between the USA and the UK. This whole episode happened in just a few hours, with the three combinations of machines held up in Despatch Area being valued at approximately £750 000.

In this particular case, although the deviation from the Terms of Reference only lasted for a few hours, it was nevertheless a deviation. Moreover the Project Engineer had no line authority to set up and carry out these tests, or to hold up the three combinations in the despatch area. However, at the time it was no one particular person's responsibility to question either the instruction from the USA or the quality of the PCBs themselves. Had there been a strong element of Quality Assurance in the company, then it would have been an automatic reaction to check these facts. It is for individuals to judge whether the Project Engineer stepped out of line; but it should be understood that there was no time to consult the senior management, and the middle managers were not willing to be accountable for decisions at this level.

9.8 CONCLUSIONS

From a Project Management aspect the actual outcome of this specific project is unimportant. To summarise briefly for completeness, little real change was made to the actual test procedures.

(1) The action concentrated effort in other areas to reduce the faults encountered during the link-up tests; that is, effort was made to find the faults at earlier stages when they would be more easily identified.
(2) The presence of a Project Engineer dedicated to highlighting problems and involving the correct people to resolve them, obviously meant that procedures ran more smoothly, and became smoother more quickly as he could push modifications and changes through.

As a consequence of the Project Engineer's work, the machine test time was reduced to 2 days without changing the procedures, and the bottleneck was overcome. The choice of Automatic Test Equipment took many months to finalise, but it was made in time to facilitate the testing of a new range of machinery as it was introduced.

There is little merit in going into more detail about the outcome of this project. However, it is of great value to draw together the key points which affected this project, not least because they will also affect other projects. These key points are listed in figure 9.3.

(a) Be clear about your line of command, who do you report to?
(b) Be clear about your objectives.
(c) Consult every interested party before and during changes in procedure and organisation, and do not just tell them what is happening but listen to them.
(d) Bring all interested parties together to resolve or prioritise conflicting objectives.
(e) You are useful because you are a neutral observer; do not become bogged down in detail, and re-read your Terms of Reference frequently to keep your efforts effective and objective.
(f) Set target dates and have progress meetings as these can resolve problems by everyone involved communicating and exchanging ideas.
(g) Your ability to mix with people, and communicate freely, will govern success.
(h) Do not over-complicate issues; in most cases common sense will generate success and identify problem areas.

Figure 9.3.   Key points for a project engineer

9.9.   DISCUSSION QUESTIONS

1. What is the difference between Quality Control and Quality Assurance? Why is Zero Defects Policy desirable?
2. Explain how the project in this case study should have been set up and controlled.

Appendix A
Discontinued Cash Flow Techniques for Project Evaluation

For projects that have to be financially viable, the cost of the project has to be recovered over a finite period. The interest rate applied to the project cost will depend on the source of the capital; money from the shareholders' funds is generally cheapest, but not always available. The shareholders' funds comprise equity (the money raised by selling shares) and profits. Interest has to be charged on shareholders' funds in order to provide for the dividend payable to shareholders. The ratio of shareholders' funds to money borrowed is the gearing ratio.

The interest rate on borrowed money may or may not be fixed in the terms of the loan. Fixed rate loans are available for medium-term or long-term borrowings (longer than a few years); while they are liable to be more expensive, they offer the advantage of making planning easier. The interest rate applied to the cost of a project may thus be based on:

(1) the interest rate for borrowed money (either a current rate or a forecast rate);
(2) the rate of return earned by the capital already invested in the business, or some similar criterion;
(3) a combination of the external and internal rates described above.

Because of interest charges, money that becomes available in the future is only worth a proportion of its current value; that is, £100 now is worth more than £100 in three years' time. Thus money in future years is discounted by a Present Value Factor, to give the Present Value. This is perhaps easier to understand from the converse process, namely compounding. Consider £100 invested at 5 per cent per annum for 5 years; the value of the investment at the end of each year is shown overleaf:

Year	Value of Investment (£)
0	100
1	105
2	110.25
3	115.76
4	121.55
5	127.63

This is expressed as:

Value of Investment at year $n = £100 \times (1+i)^n$ where i is the interest rate.

In other words, if the interest rate is 5 per cent per annum, then £115.76 in year 3 has a Present Value (PV) of £100, and the Present Value Factor is 100/115.76 or 0.864.

In general, the Present Value Factor

$$PVF = \frac{1}{(1+i)^n}$$

Values of the Present Value Factor are tabulated for different interest rates and different times in table A.1. For large values of n

$$(1+i)^n \rightarrow e^{in}$$

where i is the interest rate per time interval and n is the number of time intervals.

Since $e^{0.7} \approx 2$, this enables rapid estimation to be made of the period required for an investment to double in value; for example, it takes 14 years for an investment to double if the interest rate is 5 per cent per annum.

For project evaluation using Discounted Cash Flow (DCF), the discounting principles can be applied in two ways:

Net Present Value (NPV)
Internal Rate of Return (IRR)

The Net Present Value is calculated from the following steps:

(1) decide on the interest rate for the discounting;
(2) estimate each years net profits (that is, after tax) and multiply by the Present Value Factor to give Present Values for each year over the project life;
(3) add the Present Values from each year and subtract the net project cost to give the Net Present Value (the net project cost is the project cost after making allowance for any tax concessions or grants);
(4) calculate the Profitability Index (the ratio of the Total Present Value to the Net Project Cost).

The most desirable project appears to be the one with the highest Net Present Value. This is illustrated by the following example for Projects A and B, table A.2.

Table A.1 Present value factors for discounted cash flow calculations

Year							Annual interest rate							
	1%	2%	3%	4%	5%	6%	7%	8%	9%	10%	12%	15%	20%	25%
1	0.9901	0.9804	0.9709	0.9615	0.9524	0.9434	0.9346	0.9259	0.9174	0.9091	0.8929	0.8696	0.8333	0.8000
2	0.9803	0.9612	0.9426	0.9246	0.9070	0.8900	0.8734	0.8573	0.8417	0.8264	0.7972	0.7561	0.6944	0.6400
3	0.9706	0.9423	0.9151	0.8890	0.8638	0.8396	0.8163	0.7938	0.7722	0.7513	0.7118	0.6575	0.5787	0.5120
4	0.9610	0.9238	0.8885	0.8548	0.8227	0.7921	0.7629	0.7350	0.7084	0.6830	0.6355	0.5718	0.4823	0.4096
5	0.9515	0.9057	0.8626	0.8219	0.7835	0.7473	0.7130	0.6806	0.6499	0.6209	0.5674	0.4972	0.4019	0.3277
6	0.9420	0.8880	0.8375	0.7903	0.7462	0.7050	0.6663	0.6302	0.5963	0.5645	0.5066	0.4323	0.3349	0.2621
7	0.9327	0.8706	0.8131	0.7599	0.7107	0.6651	0.6227	0.5835	0.5470	0.5132	0.4523	0.3759	0.2791	0.2097
8	0.9235	0.8535	0.7894	0.7307	0.6768	0.6274	0.5820	0.5403	0.5019	0.4665	0.4039	0.3269	0.2326	0.1678
9	0.9143	0.8368	0.7664	0.7026	0.6446	0.5919	0.5439	0.5002	0.4604	0.4241	0.3606	0.2843	0.1938	0.1342
10	0.9053	0.8203	0.7441	0.6756	0.6139	0.5584	0.5083	0.4632	0.4224	0.3855	0.3220	0.2472	0.1615	0.1074
11	0.8963	0.8043	0.7224	0.6496	0.5847	0.5268	0.4751	0.4289	0.3875	0.3505	0.2875	0.2149	0.1346	0.0859
12	0.8874	0.7885	0.7014	0.6246	0.5568	0.4970	0.4440	0.3971	0.3555	0.3186	0.2567	0.1869	0.1122	0.0687
13	0.8787	0.7730	0.6810	0.6006	0.5303	0.4688	0.4150	0.3677	0.3262	0.2897	0.2292	0.1625	0.0935	0.0550
14	0.8700	0.7579	0.6611	0.5775	0.5051	0.4423	0.3878	0.3405	0.2992	0.2633	0.2046	0.1413	0.0779	0.0440
15	0.8613	0.7430	0.6419	0.5553	0.4810	0.4173	0.3624	0.3152	0.2745	0.2394	0.1827	0.1229	0.0649	0.0352
16	0.8528	0.7284	0.6232	0.5339	0.4581	0.3936	0.3387	0.2919	0.2519	0.2176	0.1631	0.1069	0.0541	0.0281
17	0.8444	0.7142	0.6050	0.5134	0.4363	0.3714	0.3166	0.2703	0.2311	0.1978	0.1456	0.0929	0.0451	0.0225
18	0.8360	0.7002	0.5874	0.4936	0.4155	0.3503	0.2959	0.2502	0.2120	0.1799	0.1300	0.0808	0.0376	0.0180
19	0.8277	0.6864	0.5703	0.4746	0.3957	0.3305	0.2765	0.2317	0.1945	0.1635	0.1161	0.0703	0.0313	0.0144
20	0.8195	0.6730	0.5537	0.4564	0.3769	0.3118	0.2584	0.2145	0.1784	0.1486	0.1037	0.0611	0.0261	0.0115

Table A.2 Project comparison using Net Present Values

Year no.	Present Value Factors 12 per cent	Project A		Project B	
		Net Income	Present Value	Net Income	Present Value
1	0.8929	1 000	893	2 000	1 786
2	0.7972	5 000	3 986	8 000	6 378
3	0.7118	4 000	2 847	6 000	4 271
4	0.6355	3 000	1 907	6 000	3 813
5	0.5674	2 000	1 135	6 000	3 404
6	0.5066	1 000	507	4 000	2 026
Totals		16 000	11 275	32 000	21 678
Net Project Cost			10 000		20 000
Net Present Value			1 275		1 678
Profitability Index		11 275/10 000 = 1.12		21 678/20 000 = 1.08	

Thus Project B would appear to be more attractive since it has the greater Net Present Value. However, the Profitability Index favours Project A, because here the project cost is taken into account.

The Internal Rate of Return is the interest rate that would make the project just break even—that is, give a zero Net Present Value. The internal rate of return has to be found by interpolation and, the higher the value the more desirable the project. Two iterations are shown in table A.3 for finding the Internal Rate of Return for Project A.

Further interpolation suggests an Internal Rate of Return for Project A of 16.63 per cent. When the same calculations are performed for Project B, the internal rate of return is found to be 14.81 per cent. Thus according

Table A.3

Year no.	Net Income	Present Value Factors at 16 per cent	Present Value	Present Value Factors at 16.5 per cent	Present Value
1	1 000	0.8621	862	0.8584	858
2	5 000	0.7432	3 716	0.7368	3 684
3	4 000	0.6407	2 563	0.6324	2 530
4	3 000	0.5523	1 657	0.5429	1 632
5	2 000	0.4761	952	0.4660	932
6	1 000	0.4104	410	0.4000	400
Totals	16 000		10 160		10 036
Net Project Cost			10 000		10 000
Net Present Value			160		36

to the Internal Rate of Return criterion, Project A is the most desirable since it has a higher rate of return. This is the same result as that given by the Profitability Index.

These examples were contrived so that the ratio of Total Net Income to Project Cost was the same in each case. Thus a difference is only found when the time value of money is considered by the use of discounting. In these examples no account was taken of the value of the plant at the end of its useful life. A value can be assigned and discounted by treating it as income in the final year. However, the disadvantages of a discounting system are:

(1) future net income has to be estimated;
(2) the useful life of the plant has to be estimated;
(3) in the case of the Net Present Value approach, a discounting rate has to be assumed.

The assumptions in a discounted cash flow analysis are most significant when long-term forecasts are involved. If a project has either:

(1) a short payback period, or
(2) a high ratio of total profit to capital employed, or
(3) a high annual return on investment,

then a discounted cash flow analysis is probably superfluous.

Since DCF analyses are popular with accountants, it is important for engineers to understand the technique, if only to ensure that the forecasts they make are adequate to justify a project they view as essential. Sir Francis Tombs (1984) has criticised DCF techniques for inhibiting any project that has a payback period of over five years. He suggests that this may be good for short-term results, but that long-term growth and returns are likely to suffer.

0269

Appendix B
Guidelines for Reports and Presentations

B.1 INTRODUCTION

Throughout the course of a project, engineers can expect to produce reports and presentations. It is thus essential for engineers to be able to communicate effectively using a variety of media, but especially with written reports and verbal presentations. The quality of the report or presentation is just as important as the technical content. If a judgement has to be made on the basis of a report, and the Manager cannot understand the technicalities, that will not necessarily prevent a judgement being made. Instead, the judgement will be based on a subjective assessment of the report; if the report is well written and sensibly structured with clear recommendations, then those recommendations are more likely to be followed. As always, a golden rule is—'don't save your own time by wasting your bosses' time'.

The following are some points that need to be clarified before a report or presentation can be prepared:

(1) Who is the information for?—colleagues, customers, superiors or subordinates.
(2) What is the purpose of the information?—to instruct, inform, influence, record or recommend.
(3) Plan the structure—very often the structure will be governed by in-house conventions. As a minimum a report should have an introduction, main text and conclusions; a summary is also of great use. In verbal presentations, the counterpart is 'say what you're going to say, say it, say what you've said'. As stated by Carroll (1876) 'What I tell you three times is true!'
(4) What is the scope of the presentation or report?—how much material should be covered and to what depth.

B.2   REPORTS

In order to give a good subjective impression, reports need to be well structured and well written. Poorly expressed arguments, incorrect spelling and poor grammar are all liable to prejudice the reader, regardless of the technical quality of the report.

The summary is in some ways the most important section; on the basis of reading the summary the remainder of the report may or may not be read. A busy reader will then probably turn to the introduction, conclusions and any figures. Consequently, these sections should be informative, self-contained, succinct and easy to assimilate.

B.3   PRESENTATIONS

Many aids are available for presentations; they include blackboards, whiteboards, overhead projectors, flipcharts, 35 mm transparencies, videos, films, models, samples and handouts. After establishing what is available, the most appropriate media need to be selected. Avoid using too many different aids, since this just increases the scope for mishaps. The structure of a typical presentation might be:

(1) introduction (personal and subject matter of the presentation);
(2) main content;
(3) conclusions (summary of the main points and recommendations).

The main content can be structured by using headings that will occupy 2–3 minutes of the presentation. This amount of information can usually be contained on a single transparency, or page of a flipchart. Diagrams, charts, tables and graphs can all increase the impact, but they must never be too detailed. A presentation should preferably last no longer than 20–30 minutes unless there is some organised audience activity as part of the presentation. Presentations should be as short and sharp as possible; avoid detail and technical complexity. Nothing is gained by confusing the audience, but more detail can always be supplied in response to questions (perhaps planted), or through the use of a handout.

Unless a person gives presentations regularly, he would be well advised to prepare a full script. The act of preparation helps to focus the mind, and its presence during the presentation will often mean that it is not needed. It is obviously sensible to have pages of the script corresponding directly to, say, overhead projector transparencies.

Finally, the manner in which the presentation is given is important, the aims should include:

(1) To speak clearly, varying the pace and pitch of the voice;

(2) to look at the audience and respond to them;
(3) to avoid mannerisms;
(4) to show confidence and commitment.

Once a presentation has been prepared it can be very useful to have a dress rehearsal, perhaps with some colleagues as the audience. Before a presentation it is essential to check the venue and its facilities. If any equipment is being used, it is as well to have a contingency plan should there be any malfunction.

B.4 CONCLUSIONS

Reports and presentations should both be prepared as carefully as possible, since their quality may be the basis for deciding a project's viability. In each case the reader or audience needs to be carefully identified and specifically catered for. Always aim for clarity and ease of assimilation unless the specific intention is to cause confusion.

Appendix C
Effective Working

C.1 INTRODUCTION

Effective working is essential if an individual is to achieve his full potential or perform a given job with the minimum of fuss. Perhaps many of the following comments can be regarded as common sense; if this is the case, it should not detract from their utility—common sense can be a rare commodity. There are two distinct types of working, either as an individual or as part of a team; both of these aspects are discussed here. The principal source of the following precepts is *Manage or be Managed* by Don Fuller (1963). However, to start with are some rules, in table C.1, that are expounded by Dale Carnegie (1938) in *How to Win Friends and Influence People*, a book that is probably more often quoted than read. While some of the rules may not seem appropriate, since inter-personal skills are so important, they are quoted here in full.

Table C.1   Ways for better teamwork*

TWELVE WAYS OF WINNING PEOPLE TO YOUR WAY OF THINKING

1. The only way to get the best of an argument is to avoid it.
2. Show respect for the other man's opinions. Never tell a man he is wrong.
3. If you are wrong, admit it quickly and emphatically.
4. Begin in a friendly way.
5. Get the other person saying 'yes, yes' immediately.
6. Let the other man do a great deal of the talking.
7. Let the other man feel that the idea is his.
8. Try honestly to see things from the other person's point of view.
9. Be sympathetic to the other person's ideas and desires.
10. Appeal to the nobler motives.
11. Dramatize your ideas.
12. Throw down a challenge.

Table C.1 Continued

NINE WAYS TO CHANGE PEOPLE WITHOUT GIVING OFFENCE OR
AROUSING RESENTMENT

1. Begin with praise and honest appreciation.
2. Call attention to people's mistakes indirectly.
3. Talk about your own mistakes before criticizing the other person.
4. Ask questions instead of giving direct orders.
5. Let the other man save his face.
6. Praise the slightest improvement and praise every improvement. Be 'hearty in your approbation and lavish in your praise'.
7. Give the other person a fine reputation to live up to.
8. Use encouragement. Make the fault seem easy to correct.
9. Make the other person happy about doing the thing you suggest.

SIX WAYS TO MAKE PEOPLE LIKE YOU

1. Become genuinely interested in other people.
2. Smile.
3. Remember that a man's name is to him the sweetest and most important sound in the English language.
4. Be a good listener. Encourage others to talk about themselves.
5. Talk in terms of the other man's interest.
6. Make the other person feel important—and do it sincerely.

---

* Reprinted by permission of Simon Schuster from Dale Carnegie, *How to Win Friends and Influence People*, © Dale Carnegie 1936, renewed © 1964 Donna Dale Carnegie and Dorothy Carnegie.

C.2 INDIVIDUAL WORKING

The following suggestions might help individuals to work to their full potential:

(1) recognise your own peak working time, and use this for the most critical or difficult activities—for most people this occurs in the morning;
(2) arrange for sufficient variation in your tasks so as to maximise the job interest, while avoiding fragmentation of your time through continual changes;
(3) if you have many distinct activities prepare lists, and if necessary ration or schedule time for essential activities;
(4) learn to delegate, using as many people as possible, each to his own ability;
(5) maintain a clear view of your own timetable, and avoid or reschedule activities that fragment your time away from your own place of work.

C.3 GROUP INTERACTIONS

Group interactions occur in formal occasions, such as meetings, and in less formal occasions. Since meetings can have a major impact on projects it is as well to be fully prepared, and this does not just refer to the technical aspects.

Opponents to a scheme will try and block progress by various tactics, for instance:

(1) by broadening the issues;
(2) by muddling the objectives;
(3) by raising new issues as soon as a decision approaches;
(4) by asking for definitions or clarification of issues;
(5) by showing enthusiasm but misconstruing the objectives.

There are also 'standard' forms of objection, particularly for new ideas:

(1) it's too expensive/not budgeted for;
(2) we've tried it before;
(3) we've never tried it before;
(4) the old system works, why change it for the sake of change;
(5) the unions/directors will object;
(6) has anyone else tried it, let's wait;
(7) it's too advanced/radical;
(8) it will make the present system obsolete.

By being prepared for such objections the proponents of an idea can have the answers ready. Furthermore, during the course of meetings, the proponents of a project can adopt a strategy to help achieve their aim by:

(1) making the objectives clear;
(2) focusing attention on the main issues;
(3) preparing their case thoroughly;
(4) ensuring that the meeting does not become sidetracked—it can be very useful to have and follow an agenda.

Less formal meetings can be equally important, and the same principles apply. Care is always needed to ensure that people do not feel that their ideas are being either ignored or ridiculed. Equally important is the avoidance of confrontation (especially in front of others), since this can lead to someone's loss of face.

Delegation can also be a very sensitive issue, since the amount of supervision and support can be critical. If too much time is spent defining the problem, no time will be saved by delegation; equally, if too little time is spent the job may be done poorly. Too much support shows a lack of confidence, while too little support would demonstrate a lack of interest. There can be no universal solution, but a clear and logical statement of the problem, with help always given when requested, should make it easier. If the task is carried out successfully, the full credit should be given; but if a mistake occurs, then the blame should be shared. Criticising with the benefit of hindsight is never likely to be constructive.

C.4 DECISION MAKING

Many types of decision have to be made and whether they are irreversible or reversible, major or small, unique or repetitive, or new or old will influence how the decision is made. Obviously, most care is needed for an irreversible major decision, for which there is no previous experience. It may be necessary to consider all the implications of a decision, in which case the decision can be influenced by attempting to minimise any risk. Always be prepared to ask for advice, and accept the fact that mistakes are likely to be made: 'Someone who never makes mistakes never learns or does anything'.

In decision making it is wise to distinguish between should, must, can and will (******); the differences in emphasis may be subtle, but they are also significant.

Who	******	make the decision?
Who	******	advise on the decision?
Who	******	be involved in making the decision?
Who	******	be informed of the decision?
What	******	be decided?
When	******	the decision be made/announced/implemented?

(****** represents: should, must, can or will)

Decision making depends on having sound information, and there are many traps to be fallen into:

(1) make sure you ask the right question;
(2) look out for the prejudices of people giving 'information';
(3) look out for non-sequiturs and other forms of fake logic;
(4) ensure that samples are adequate, and be wary of 'averages'.

Appendix D
The Laws of Project Engineering

The following material is not to be taken too seriously, but all the same there are some serious messages to be found. To quote from W. S. Gilbert:

'Oh winnow all my folly and you'll find
a grain or two of truth among the chaff'

Like all the best subjects (namely thermodynamics), there are two principal laws. The first law is that if something can go wrong on a project it will. The second law is difficult to express in a precise form but, in general terms, it is that things tend to go from bad to worse. There are also many corollaries to this second law:

Nothing is as easy as it looks.

Everything takes longer than you think.

Sooner or later the worst possible set of circumstances is bound to occur.

It is impossible to make anything foolproof, because fools are so ingenious.

Firmness of delivery dates is inversely proportional to the tightness of the schedule.

An object or bit of information most needed will be the least available.

In any collection of data, the figure most obviously correct beyond all need of checking is the mistake.

If there is a possibility of several things going wrong, the one that will cause most disruption will go wrong.

Failures will not occur until after the final inspection.

In addition, any project associated with a computer will be subject to the following corollaries:

Any given program, when running, is obsolete.

Any given program will expand to fill all the available memory.

Undetectable errors are infinite in their variety.

A computer program never does what you want, only what you tell it to.

Computers are unreliable, but humans are worse.

Adding manpower to a late software project makes it later.

The applied law of entomology: there is always one more bug.

This material has been adapted from *Murphy's Law* and *Murphy's Law Book Two*, both by Arthur Bloch, Price, Stern & Sloan, 1977 and 1980 respectively.

Appendix E
The North West Corner Algorithm

When the locations and capacities of a series of facilities that supply a range of customers need to be optimised, it is necessary to use a linear programming technique. The following is a very simple example, since it considers only one combination of factories and warehouses, and only a single product is considered. For a complete solution the distribution from the warehouses also needs to be evaluated, and other combinations of factories, warehouses and consumers would need consideration.

Example

Three factories produce sugar at the rate of

   1 ≡ 75 tonnes/day
   2 ≡ 50 tonnes/day
   3 ≡ 30 tonnes/day

and the three proposed warehouses would distribute sugar at the rate of

   A ≡ 25 tonnes/day
   B ≡ 60 tonnes/day
   C ≡ 70 tonnes/day

Fortunately, supply and demand are balanced in the example; if there was an imbalance, a dummy warehouse would have to be added. The cost of transport per unit mass between the factories and warehouses is most readily expressed in a matrix (figure E.1).

The first transportation tableau can now be constructed using the transport cost matrix, the factory production rates and the warehouse demand rates (figure E.2).

Warehouses / Factories	A	B	C
1	3	2	6
2	5	4	1
3	1	6	3

Figure E.1   Transport cost matrix (C_{ij})

Warehouses $(j = A, B, C)$ / Factories $(i = 1, 2, 3)$	A	B	C	Daily factory output (a_i)
1	X_{ij} 3	X_{ij} 2	X_{ij} 6	$a_1 = 75$
2	X_{ij} 5	X_{ij} 4	X_{ij} 1	$a_2 = 50$
3	X_{ij} 1	X_{ij} 6	X_{ij} 3	$a_3 = 30$
Warehouse daily requirements, (b_j)	$b_A = 25$	$b_B = 60$	$b_C = 70$	

Figure E.2   Transportation tableau

The object is to satisfy the distribution requirements and to minimise the transportation costs. A suitable transportation algorithm is the *north west corner* rule; this is shown as a flow diagram in figure E.3.

Applying this procedure leads to the following steps, which create figure E.4:

(1) start at A1;

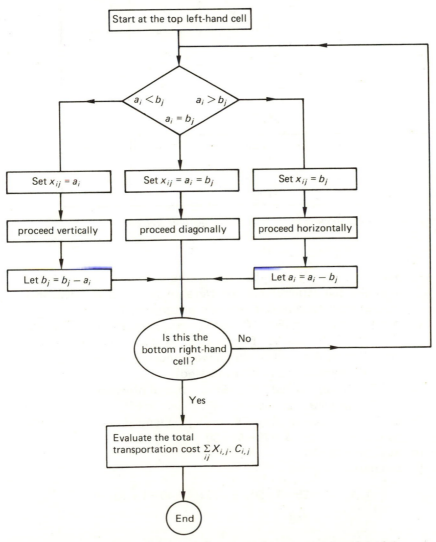

Figure E.3   Flow diagram for the north west corner transportation algorithm

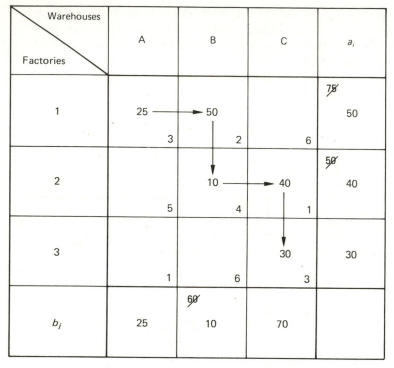

Figure E.4   First solution to the transport allocation problem

(2) since $a_i > b_j$ (75 > 25), set $x_{ij} = b_j$ and proceed horizontally to B1; let $a_i = a_i - b_j = 75 - 25 = 50$ (that is, the output of factory A has met the full requirement for warehouse 1 and the available output from factory A is reduced accordingly);

(3) at B1, since $a_i < b_j$ (50 < 60), set $x_{ij} = a_i$ and proceed vertically to B2; let $b_j = b_j - a_i = 60 - 50 = 10$ (that is, the remaining output from factory A is insufficient for warehouse 2, and the outstanding requirement has to be evaluated);

(4) at B2, since $a_i > b_j$ (50 > 10), set $x_{ij} = b_j$ and proceed horizontally to C2; let $a_i = a_i - b_j = 50 - 10 = 40$;

(5) at C2, since $a_i < b_j$ (40 < 70), set $x_{ij} = a_i$ and proceed vertically to C3; let $b_j = b_j - a_i = 70 - 40 = 30$

(6) at C_3, $a_i = b_j = 30$;

(7) evaluate

$$\sum_{i,j} X_{ij} \cdot C_{ij} = 25 \times 3 + 50 \times 2 + 10 \times 4 + 40 \times 1 + 30 \times 3$$

$$= 345$$

This solution is unlikely to be the best possible; to optimise

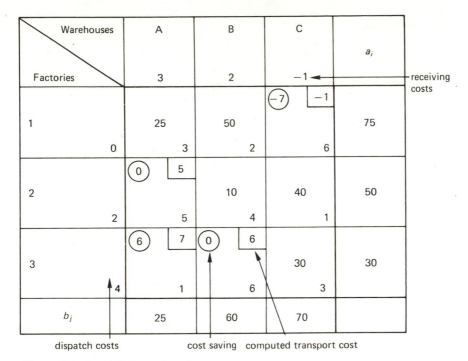

Figure E.5   Use of dispatch and receiving costs

the solution the concepts of dispatch and receiving costs are adopted. The transportation cost between each location is equal to the sum of the relevant dispatch and receiving costs. Starting with factory 1 the dispatch cost can be set to any value, and in this instance zero is adopted. Thus the receiving cost at warehouse A is 3.

Using only the routes adopted in the first solution (figure E.4), the receiving and dispatch costs can be evaluated in the following order as:

receiving at B	2
dispatch from 2	2
receiving at C	−1
dispatch from 3	4

The negative receiving cost at C does not matter since the dispatch cost from factory 1 was arbitrarily set to zero. The dispatch and receiving costs are shown on figure E.5, along with the computed costs for the routes not used; these computed costs are shown in the top right-hand corner of the relevant boxes.

The difference between the computed cost and the actual cost is the transport cost that could be saved by adopting that route;

Warehouses / Factories	A	B	C	a_i
1	$25 - Q$ 3	$50 + Q$ 2	 6	75
2	 5	$10 - Q$ 4	$40 + Q$ 1	50
3	Q 1	 6	$30 - Q$ 3	30
b_i	25	60	70	

Figure E.6   Re-allocation of quantity Q

such savings are circled in the top left-hand corner of each relevant box. For example, the computed cost of the route 3 to A is 7 while the actual cost is 1, a potential saving of 6. Since this is the largest potential saving it would be advantageous for some sugar to travel along this route.

In figure E.6 let the quantity Q travel along route 3 to A. The quantities travelling from 1 to A and from 3 to C both need to be reduced by Q, so that the effect of Q on the rows and columns balances out. This principle has been applied to the remainder of figure E.6.

Since negative quantities of the product are not permissible, the largest value that Q can take is 10, and the revised allocations for each route are shown in figure E.7. At this stage the dispatch and receiving costs are determined for the routes in current use. Again setting the dispatch cost from factory 1 to zero, the costs are evaluated in the following order:

receiving at A	3
dispatch from 3	−2
receiving at B	2
receiving at C	5
dispatch from 2	−4

Warehouses / Factories	A	B	C	a_i
(costs)	3	2	5	
1 0	15 _(cost 3)_	60 _(cost 2)_	(−3) 3	75
2 −4	(−4) 1 _(cost 3)_	(−2) 2 _(cost 2)_ 0	50 _(cost 6)_	50
3 −2	10 _(cost 5)_	(−6) 0 _(cost 4)_	20 _(cost 1)_	30
b_j	25	60	70	

Figure E.7 *Optimum route allocation*

These costs are shown in figure E.7, along with the computed costs for the routes not used, and the possible savings. Since the possible savings are negative, the optimum route allocation has been found, and the total transport cost $(\sum_{ij} X_{i,j} \ C_{i,j})$ is 285.

When a dummy warehouse has been introduced to absorb a potential over-production, the transportation costs should always be set to zero, since in practice the factory output to the dummy warehouse would not be produced. In conclusion, it must be remembered that transport costs are not the only consideration in the siting of a new facility.

References

R. Bell (1984), 'Programmable electronic systems: the safety aspects', *Process Engineering*, April 1984, pp. 21-4.

D. W. Birchall and R. Newcombe (1985), 'Developing project manager skills', *Metals and Minerals*, July 1985.

R. H. W. Brook (1974), 'How safe is safe?', *CME*, Vol. 21, No. 10, pp. 75-8.

A. J. Brown (1984), 'Financing major export contracts', *Proc. I. Mech. E.*, Vol. 198B, No. 10 pp. 203-6.

BS 5034 (1988) *Code of Practice for Safety Machinery*, British Standards Institution, London.

BS 5750 (1979), *Quality Systems, Parts 1 to 6*, British Standards Institution, London (1979-1981).

C. A. Carnall (1984), *Organization Behaviour and Industrial Relations*, The Management College, Henley.

D. Carnegie (1938), *How to Win Friends and Influence People*, Worlds Work, Tadworth.

L. Carroll (1876), *The Hunting of the Snark*, London.

J. Child (1984), *Organisation—A Guide to Problems and Practice*, Harper and Row, London.

E. B. Cowell (1982), 'The integration of project time control and project cost control', *Proc. I. Mech. E.*, Vol. 196, No. 28, pp. 313-16.

M. Dalton (1959), *Men Who Manage*, Wiley, London.

D. Davies, T. Banfield and R. Steahan (1976), *The Humane Technologist*, Oxford University Press.

S. M. Davis and P. R. Lawrence (eds) (1977), *Matrix*, Addison-Wesley, London.

B. Fischoff, S. Lichtenstien, P. Slovac, S. L. Derby and R. L. Keeney (1981), *Acceptable Risk*, Cambridge University Press.

D. Fuller (1963), *Manage or Be Managed*, Industrial Education Institute, Boston, Massachusetts.

J. R. Galbraith (1971), *Matrix Organisation Designs*, Business Horizons, February 1971.

C. B. Handy (1981), *Understanding Organisations*, Penguin, London.

F. L. Harrison (1985), *Advanced Project Management*, Gower Press, Farnborough.

B. Harvey (1983), 'Point of view', *The Safety Practitioner*, October 1983, pp. 18-21.

A. Jay (1967), *Management and Machiavelli*, Hodder and Stoughton, London.

R. J. Kingsley, M. Neale and E. Schwartz (1969), 'Commissioning of medium-scale chemical process plant', *Proc. I. Mech. E.*, Vol. 183, Pt I, No. 11, pp. 205–18.

T. Kletz (1979), 'Safety in numbers', *Focus on Engineering* (ICI), Hobsons.

K. Knight (1977), *Matrix Management*, Gower Press, Farnborough.

M. C. Lawrence (1978), 'Hearing protection', *Electronics and Power*, January 1978, pp. 58–61.

D. Lock (1971), 'Scheduling industrial projects', *CME*, Vol. 18, No. 9, pp. 304–10.

D. Lock (1977), *Project Management*, 2nd edition, Gower Press, Farnborough.

P. W. G. Morris (1987), 'The everyday challenges of major projects', *CME*, Vol. 35, No. 3, pp. 33–8.

J. D. Murray (1984), 'The design, construction and installation of a continuous casting plant for the production of special steel billets at Stocksbridge Works,' *Proc. I. Mech. E.*, Vol. 198, No. 42.

R. Muther (1973), 'Planning for new facilities and modifications to existing facilities', from B. T. Lewis and J. P. Marron (eds), *Facilities and Plant Engineering Handbook*, McGraw-Hill, New York.

J. Oddey (1981), 'Back to basics', *CME*, Vol. 28, No. 11, pp. 52–7.

R. Oxley and J. Poskitt (1980), *Management Techniques applied to the Construction Industry*, 3rd edition, Granada, London.

Oyez (1978), *High Risk Exposures*, Oyez Intelligence Reports, Oyez Publishing Ltd, London.

C. N. Parkinson (1981), *The Law*, Penguin, London.

N. Percival (1983), 'Developments in machine guarding and protection systems', *The Safety Practitioner*, August 1983, pp. 8–11.

K. K. Pillai 1983), 'Combined cycle power plant utilizing pressurized fluidized-bed combustors, Paper C72/83, *Combustion in Engineering*, Vol. II, I. Mech. E. Conference Publications, MEP, London.

S. Potter (1987), 'Why did the APT Fail?', *Modern Railways*, January 1987, pp. 30–4.

M. Smith (1982), Noise control—the good and the bad news', *CME*, Vol. 29, No. 10, pp. 26–8.

A. Spinks (1974), 'Escalating entrance fees for new drugs', *ICI Magazine*, Vol. 52 No. 411, pp. 122–8.

L. C. Stuckenbruck (1981), *The Implementation of Project Management; The Professional's Handbook*, Addison-Wesley, London.

Sir Francis Tombs (1984), 'The way ahead for manufacturing industry', *Proc. I. Mech. E.*, Vol. 198B, No. 14.

D. J. Warby (1984), 'Preparing the offer', *Proc. I. Mech. E.*, Vol. 198B, No. 10, pp. 195–201.

S. H. Wearne (1979), 'A review of reports of failure', *Proc. I. Mech. E.*, Vol. 193, No. 20, pp. 125–30.

S. H. Wearne (1984), 'Contractual responsibilities for the design of an engineering plant: a survey of practice and problems', *Proc. I. Mech. E.*, Vol. 198B, No. 6, pp. 97–108.

R. Wild (1984), *Production and Operations Management*, 3rd edition, Holt, Rinehart and Winston, London.

Bibliography

GENERAL READING

F. L. Harrison, E. A. Stallworthy and O. P. Kharbanda (1983), *Total Project Management*, Gower Press, Farnborough. Institution of Chemical Engineers Publications, Rugby.

O. P. Kharbanda and E. A. Stallworthy (1983), *How to Learn from Project Disasters; True Life Stories with a Moral for Management*, Gower Press, Farnborough.

T. Kidder (1981), *The Soul of a New Machine*, Penguin, London.

S. H. Wearne (ed.) (1984), *Control of Engineering Projects*, Arnold, London.

CONTRACT LAW

G. C. Cheshire and C. H. S. Fifoot (1981), *Law of Contract*, Butterworths, London.

F. R. Davies (1981), *Contract*, Sweet & Maxwell, London.

R. M. Goode (1982), *Commercial Law*, Penguin, London.

R. Lowe (1983), *Commercial Law*, Sweet & Maxwell, London.

G. F. Woodroffe (1982), *Goods and Services—The New Law*, Sweet & Maxwell, London.

HEALTH AND SAFETY

There is an enormous literature on health and safety, particularly that published by the Health and Safety Commission and the Health and Safety Executive. Those working in particular industries should always ensure that they have access at least to those publications that relate to their particular industry. Details from:

- The Health and Safety Commision, St Hugh's House, Stanley Precinct, Bootle, Merseyside, L20 3QY (Tel. 051-951-4223).
- The Health and Safety Executive, 25 Chapel Street, London, NW1 5DT (Tel. 01-262-3277).

T. A. Kletz (1985), *An Engineer's View of Human Error, I. Chem. E.*, Rugby.
For introductions to the law, see for example:
R. Benedictus (1980), *Safety Representatives*, Sweet & Maxwell, London.
P. Rowe (1980), *Health and Safety*, Sweet & Maxwell, London.
N. Selwyn (1982), *Law of Health and Safety at Work*, Butterworths, London.

The legal practitioner's bible is:
M. Goodman (ed.), *Encyclopedia of Health and Safety at Work*, Sweet & Maxwell, London (looseleaf, regularly updated).

In addition, there are frequent accounts of safety law either generally or describing new developments, in professional and trade journals which assist in keeping the project engineer up to date.

COMPUTING PROJECTS

F. P. Brookes Jr (1975), *The Mythical Man-Month*, Addison-Wesley, New York.
H. D. Covvey and N. H. MacAlister (1980), *Computer Consciousness*, Addison-Wesley, New York.

REPORT WRITING

B. M. Cooper (1964), *Writing Technical Reports*, Penguin, London.

CAD/CAM

A. J. Medland and P. Burnett (1986), *CADCAM in Practice*, Kogan Page, London.

Index